SpringerBriefs in Mathematics

SpringerBriefs in Mathematics showcases expositions in all areas of mathematics and applied mathematics. Manuscripts presenting new results or a single new result in a classical field, new field, or an emerging topic, applications, or bridges between new results and already published works, are encouraged. The series is intended for mathematicians and applied mathematicians.

More information about this series at http://www.springer.com/series/10030

SBMAC SpringerBriefs

The **SBMAC SpringerBriefs** series publishes relevant contributions in the fields of applied and computational mathematics, mathematics, scientific computing, and related areas. Featuring compact volumes of 50 to 125 pages, the series covers a range of content from professional to academic.

The Sociedade Brasileira de Matemática Aplicada e Computacional (Brazilian Society of Computational and Applied Mathematics, SBMAC) is a professional association focused on computational and industrial applied mathematics. The society is active in furthering the development of mathematics and its applications in scientific, technological, and industrial fields. The SBMAC has helped to develop the applications of mathematics in science, technology, and industry, to encourage the development and implementation of effective methods and mathematical techniques for the benefit of science and technology, and to promote the exchange of ideas and information between the diverse areas of application.

http://www.sbmac.org.br/

Rosana Sueli da Motta Jafelice
Ana Maria Amarillo Bertone

Biological Models via Interval Type-2 Fuzzy Sets

 Springer

Rosana Sueli da Motta Jafelice
Faculty of Mathematics
Federal University of Uberlândia
Uberlândia
Minas Gerais, Brazil

Ana Maria Amarillo Bertone
Faculty of Mathematics
Federal University of Uberlândia
Uberlândia
Minas Gerais, Brazil

ISSN 2191-8198   ISSN 2191-8201 (electronic)
SpringerBriefs in Mathematics
ISBN 978-3-030-64529-8   ISBN 978-3-030-64530-4 (eBook)
https://doi.org/10.1007/978-3-030-64530-4

Mathematics Subject Classification: 03E72, 92-10, 92B05

This Springer imprint is published by the registered company Springer Nature Switzerland AG
The registered company address is: Gewerbestrasse 11, 6330 Cham, Switzerland

The authors dedicate this book to all the people who have contributed to the Biomathematics-Fuzzy Set Theory partnership.

Rosana dedicates this book to God, to her love, Luís Fernando, and to her sons, Mateus and Lucas.

Ana María dedicates this book to her husband, Craig, her daily joy, her "reason for being."

Preface

After several years of research and academic work, focused on modeling biological phenomena and developed with undergraduate and graduate students, the advantages of the interval type-2 fuzzy sets methodology have been verified. Noticing the lack of books about biological models using the interval type-2 fuzzy sets, the authors have explored the potential coverage in the field of this theory. This has been the foundation for this book, initially in the form of a mini-course, presented in the IV Brazilian Congress of Fuzzy Systems in 2016. After this event, the investigation continued, reviewing the theory from more recent papers and, as a consequence, improving the notes of mini-course and developing a better understanding of the value of interval type-2 fuzzy theory. In fact, the accuracy of the technique compared with the counterpart of type-1 has been measured. When no comparison with type-1 fuzzy technique is done, a range is obtained that includes other outputs of the model, extending the way to provide the approximation of the solution.

This background establishes the aim of this book: introduce applications involving biological models solved through interval type-2 fuzzy techniques. The emphasis is to use computational means in the applications, along with theoretical support. The target audience is graduate students in mathematics and related fields, professionals, researchers, or the general public who wish to grasp the topics and methods of this innovative area. To this end, it is recommended to have a background in classical fuzzy set theory, specifically type-1 fuzzy sets theory.

The highlights of the book are as follows:

- A chapter dedicated to explaining, didactically, the basic concepts of the interval type-2 fuzzy theory, with examples and challenging questions that will help the reader to understand the nature of this topic.
- Six biological applications using the interval type-2 fuzzy theory, along with their own contributions, explanations, and suggestions for further development. Highlighted is the methodology applied by diverse mathematical tools: cellular automaton, partially fuzzy (p-fuzzy) systems, and ordinary differential equations.
- The book encompasses the basic information and advanced concepts, both required to develop new applications for biological modeling.

The authors acknowledge Prof. Luiz Mariano Carvalho for the encouragement to elaborate this book, Craig Snively and Leslie Snively for English text proofing, anonymous reviewers, and Springer's editors and staff for their assistance in preparing the manuscript.

Uberlândia, Brazil   Rosana Sueli da Motta Jafelice
Uberlândia, Brazil   Ana Maria Amarillo Bertone

Contents

List of Figures

List of Tables

Chapter 1
Introduction

> ... *type-2 fuzzy sets can capture two kind of linguistic*
> *uncertainties simultaneously (the uncertainty of individual and*
> *uncertainties of a group about a word) whereas type-1*
> *cannot... [29]*
>
> Jerry M. Mendel (2015)

Two important areas of mathematics are related by modeling purposes: *Biomathematics* and *Fuzzy Sets Theory*. Both have a point in common in their history of development: a respected background in formal theory and a wide range of applications, both attained in the twentieth century. That is, both areas are new scientific tendencies in comparison to those mathematical areas with centuries of discussion and establishment.

Biomathematics refers to a branch of biology that concentrates its goals on building mathematical models to describe and solve biological problems. Traditionally, a model of a biological system was converted into a system of equations that govern the phenomenon and is capable to interpret its dynamics. From there, the equations' solution is built, either by analytic or numerical means, with the goal of responding to the question of how the biological system behaves and which features are confirmed by the modeling procedure. Classic problems, such as population growth, predator–prey interactions, and epidemics models, indicated historically the origin of the area. In this sense, the application of mathematical principles to biological processes harken back to the twelfth century, when Leonardo Fibonacci [35] used a numerical series to describe a growing population of rabbits. Moreover, in the eighteenth century, Daniel Bernoulli [7] studied the effects of smallpox on the human population, Thomas Malthus [23] and Pierre Verhulst [41] studied the growth of the human population, given, as a result, the curve's shape of the phenomenon. These classic population models have been analyzed from another point of view: the Verhulst model in Sect. 2.1.3 and a suggestion for a project to be expanded in Sect. 4.4; the Malthus model is explored in Sect. 3.5.

© The Author(s), under exclusive license to Springer Nature Switzerland AG 2021
R. S. da Motta Jafelice, A. M. A. Bertone, *Biological Models via Interval Type-2
Fuzzy Sets*, SpringerBriefs in Mathematics,
https://doi.org/10.1007/978-3-030-64530-4_1

In the early twentieth century, the Lotka–Volterra predator–prey model was proposed. Alfred Lotka used the equations to analyze predator–prey interactions in his book on biomathematics [22]. The same set of equations were studied by Vito Volterra, who developed his model independently from Lotka, as many famous results are found apart from brilliants minds. Highlighted in this book, the Lotka–Volterra predator–prey model is explored to provide ideas for a project to be developed in Sect. 4.5.

After the Spanish flu epidemic, the work developed by William Kermack and Anderson McKendrick [21] is considered one of the first formal models on epidemics. Currently named as SIR model is utilized in Sect. 3.6.2 to analyze the 2020 epidemic of a new coronavirus from a fuzzy perspective.

The emergence of new mathematical and computational tools, defining the characteristics of biological systems is becoming more complex and detailed. One of the complexities implicit in almost all biological systems is the uncertainty or vagueness; that comes from the nature of the phenomenon, the information of the literature, the interpretation of experts, and the collected data, among other reasons. Observing this feature is how the fuzzy sets approaches have come to aid the modeling and accuracy of its results, despite the fuzziness of their denomination. At the end, the fuzzy set theory has changed the point of view of the interpretation, having the aim of portraying the phenomenon dynamics in a more realistic way. Modeling with words, establishing rule base, and obtaining accurate mathematical results by defuzzification methods are, among other advantages, the innovative tools that the investigators of the area have in their toolbox today.

Fuzzy Logic was created by Lotfi Zadeh, professor emeritus of computer science at the University of California—Berkeley, where he proposed an alternative logic to the classical Boolean one. Zadeh's first paper in this new area of knowledge was released in 1965 [42]. Only 10 years after, Zadeh introduced type-2 fuzzy sets in 1975 [44], which extended Zadeh's original fuzzy set theory, currently known as type-1 fuzzy sets. Around the same time, control systems were built based on the fuzzy logic. As Zadeh pointed out in his paper [45]: "Fuzzy if-then rules of this type were employed by Mamdani and Assilianin 1974 in their seminal work on the fuzzy control of a steam engine." In fact, Ebrahim Mamdani and his PhD student Sedrak Assilian [24] had built an engine with a little boiler trying to control it automatically [27]. Another important historical contribution is due to Michio Sugeno who first introduced fuzzy measures and the Sugeno Integrals [38]. Afterwards, a joint work with Tomohiro Takagi [40] deployed an inference system for automatically controlling a car. For many years, this inference method was named as Takagi–Sugeno (TS). As described in Sugeno's biography: "The groundbreaking work has had a tremendous impact on fuzzy control researchers and has impacted applications such as home appliances, automobiles and process control. . . [8]." Recently, the TS inference method is more commonly known as Takagi–Sugeno–Kang (TSK) due to the joint work of Sugeno and Geun-Taek Kang [39] related to the structure identification of fuzzy models.

In the 1990s, a great advancement in the fuzzy modeling was done by Jyh-Shing Jang with his an Adaptive Network-based Fuzzy Inference System (ANFIS), based on TSK fuzzy inference method . Integrating two powerful tools, neural networks and fuzzy logic, the ANFIS method brought the advantages to capture the benefits of both in a single framework. As Jang clearly expressed in his article [17] "by employing fuzzy if-then rules, a fuzzy inference system can express the qualitative aspect of human reasoning without using any precise mathematical models of the system." The use of artificial intelligence is reinforced in Jang's work [18]. The ANFIS technique is utilized throughout the book as a means of obtaining components for the inference method.

Fifty-five years have passed since the type-1 fuzzy theory was launched, and forty five for the type-2 counterpart, both running on a similar path in terms of acceptance by the scientific community. There was some resistance early on until a flood of applications and academic papers begun to appear. Jerry Mendel, pioneering in the rescue of the Zadeh's type-2 fuzzy sets theory, introduced his first paper in the matter in 1998 [19], 23 years after Zadeh's first paper in the subject. Recently, Mendel has asked the following question [30]: "Why did Zadeh invent type-2 fuzzy sets?". During the 1960s and early 1970s, Zadeh was under attack for his work, about the real uncertainty nature of a fuzzy set. Mendel's answer is "I can only speculate that perhaps a type-2 fuzzy set was a response to this attack." Furthermore, Mendel emphasizes the theoretical and applicability contribution of type-2 in the following words:

> When unpredictability is present, [31, p. 19] most of us have no problem with using probability models and analyses from the very beginning; hence, when membership function uncertainties (e.g., due to linguistic or partition end-point uncertainties) are present, one should have no problem with using type-2 fuzzy set models and analyses from the very beginning. Some may ask: "How much membership function uncertainties must be present before type-2 fuzzy sets should be used?". Maybe, in the early days of probability, a similar question was asked; however, it no longer seems to be relevant. When randomness is suspected, probability is used. So when membership function uncertainties are suspected, type-2 fuzzy sets should be used.

The crucial difference between these two mathematical concepts is that the type-1 fuzzy set has as membership grade a single value, and the type-2 fuzzy set reveals the membership grade in two stages, the primary and secondary membership functions. When the secondary is constant, the result is called an interval type-2 fuzzy set; otherwise, it is called a general type-2 fuzzy set [28]. In 1998, after a decade, the work of Nilesh Karnik and Jerry Mendel [19, 20] appeared, with the presentation of the complete theory of the interval type-2 Fuzzy Rule-Based System (FRBS), including the type reducer as a previous stage of the defuzzification method. The execution of the interval type-2 FRBS, containing the reducer component, had a very high computational cost. Dongrui Wu and Jerry Mendel [31] developed computational algorithms that improved this negative aspect.

Another important contribution for the area was the Interval Type-2 Fuzzy Logic Toolbox [5] developed in 2007 by J. R. Castro, O. Castillo, and L. G. Martínez. This toolbox is a set of functions for building interval type-2 FRBS, utilized in the

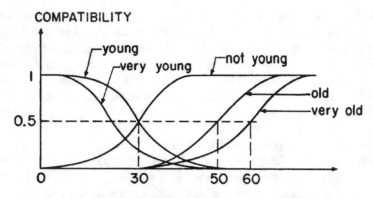

Fig. 1.1 Drawing of an interval type-2 fuzzy linguistic term. Digital extracted from the original paper [43, 44]

applications of this book, and being kindly provided to the authors for Oscar Castillo for its use.

Without doubt, Mendel has embraced the type-2 fuzzy theory and found a large community of followers, which has enriched the area in both theory and applications. In particular, the interval type-2 fuzzy theory has its own evolution that demonstrates the interest of an important group in the scientific community on the topic. In fact, from the simple drawing in Zadeh's chapter [43], where a graph of this type of mathematical object is shown (see Fig. 1.1), the interval type-2 fuzzy sets theory has paved a solid path summarized in the explosion of a number of works, publications, citations, and patents as illustrated in Fig. 1.2.

Presented in the illustration of Fig. 1.1 are the terms young and old, whose meaning might be defined by their respective compatibility functions μ_{young} and μ_{old}. Apart from these functions, other compatibility functions can be $\mu_{\text{very young}}$, $\mu_{\text{more or less young}}$, and $\mu_{\text{not very young}}$.

The results shown in the statistics of Fig. 1.2 have been obtained using the keyword "interval type-2 fuzzy" in the portal Google Scholar [9] from 1991 throughout 2020 (year in progress). Observe that from 1991 to 2005 the number of works in this particular subject is less than one hundred, and after 2005, there was a tremendous growth as shown in Fig. 1.2.

The relationship between Biomathematics and Fuzzy Sets Theory has been demonstrated in the works of Elie Sanchez and Robert Bartolin [33, 34], avant-garde of the connection between uncertainty and medical phenomena. Particularly, in their work of 1982 [1] they highlighted that "the theory of fuzzy sets and fuzzy logic is approached as a diagnostic aid for practitioners and biologists," reinforcing the close link between medical modeling and fuzzy logic.

Referring in particular to Brazil, one of the pioneers in the area of Biomathematics is Rodney Bassanezi, who collaborated for the implementation of this

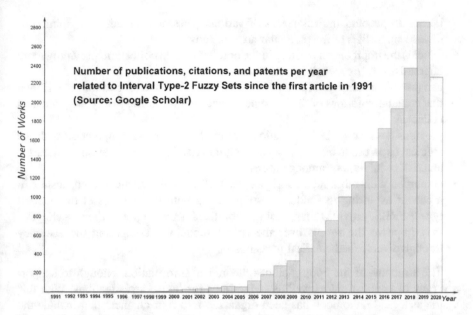

Fig. 1.2 In the bar chart is shown the number of works since 1991 up to 2020, year in progress. Source: Google Scholar

research area at the State University of Campinas [2, 3]. Many scientific publications by this researcher and collaborators have been carried out with the combination of biological and natural models along with the fuzzy theory. Applications in differential equations with fuzzy parameters have been developed in models of Human Immunodeficiency Virus (HIV) dynamics, population dispersion, heat, wave, and Poisson equations, among others [4, 11, 13, 15]. Another computational tool used to simulate biological phenomena artificially is the Cellular Automaton, which has been modeled in conjunction with fuzzy sets, in both type-1 and interval type-2 [12, 14, 16]. Cellular automata with parameters obtained from interval type-2 fuzzy sets can be found in Sect. 3.4. Theoretical work has also been a highlight in this area, for example, the studies of stationary points p-fuzzy dynamical systems [6, 36, 37]. Presented in Sect. 2.1.3, Sect. 3.3, and Sect. 3.5 are applications of the p-fuzzy system utilizing the interval type-2 fuzzy modeling. Other Brazilian research centers have developed works that related the modeling of biological phenomena and the theory of fuzzy sets [25, 26, 32].

In view of the previous considerations and as remarked in the Preface, after many year of research, workshops, and courses presented in conferences [10], the combination of Biomathematics and Interval Type-2 Fuzzy Theory has been fruitful. As a consequence, the goal of this book is to introduce the reader to this successful partnership. In fact, in the applications that have been proposed in this book, many advances have been detected in modeling the uncertainty of the biological phenomena, namely, this methodology:

- takes into account the vagueness of various types and includes the opinion of diverse specialists of the particular area in study;
- handles the imprecision existing in the practice of data collection, providing more alternatives for its model;
- can provide a range of alternatives for the model output, for example, a ribbon that contains solutions of diverse type, as deterministic, fuzzy of type-1, and data collected;
- provides a wider set of information that might be used in conjunction with the medical expert to avoid any risks of drug intoxication, identification of diseases, and risk of infections, among others;
- combined with other methodologies, such as the cellular automaton, results in a benefit exploring its randomness that, along with the features of the interval type-2 FRBS, provides a range of possibilities for the output of the modeling;
- may improve the results over the type-1 modeling, comparing the accuracy through diverse mathematical measures.

The structure of the book includes theoretical information, enough to support a variety of applications as well as suggestions for future investigations. With this vision as a guide, the book has been organized into four chapters that include the current chapter as an Introduction. Chap. 2 is a tour to the fuzzy theory of both types, with examples and exercises, all chosen for better guidance of the topics in the next chapters. In Chap. 3, interval type-2 Fuzzy Rule-Based System Applications are detailed in topics of the biological field models such as: pharmacological in Sect. 3.1, prediction of prostate cancer stage in Sect. 3.2, epidemiological disease caused by HIV in Sect. 3.3 and in Sect. 3.4, population growth in Sect. 3.5, and the epidemic caused by a new coronavirus in Sect. 3.6. In order to achieve the goal of the book, the reader is encouraged to complete the challenges proposed in Chap. 4 for five scientific projects.

Knowing that linguistic terms are difficult to model, especially the ones that are more vague such as "strongly low" and "slightly hot," the authors hope that the book's content awakens in the target audience's

"a curiosity *slightly* intense, but *strongly* pleasant."

References

1. Bartolin, R., Bouvenot, G., Soula, G., Sanchez, E.: The fuzzy set theory as a biomedical diagnostic aid. Sem. Hop. **58**(22), 1361–1365 (1982)
2. Bassanezi, R.C., Barros, L.C.: A simple model of life expectancy with subjective parameters. Kybernetes **24**(7), 57–62 (1995)
3. Bassanezi, R., Roman, H.: Relaciones fuzzy: Optimización de diagnóstico médico. Technical report, IMECC, UNICAMP, Campinas, Brazil (1989, in Spanish)
4. Bertone, A., Jafelice, R., Barros, L., Bassanezi, R.: On fuzzy solutions for partial differential equations. Fuzzy Sets Syst. **219**, 68–80 (2013)

5. Castro, J.R., Castillo, O., Martínez, L.G.: Interval type-2 fuzzy logic toolbox. Eng. Lett. **15**(1) (2007)
6. Cecconello, M.S.: Modelagem alternativa para dinâmica populacional: sistemas dinâmicos fuzzy. Master's thesis, IMECC-UNICAMP, Campinas, Brazil (2006, in Portuguese)
7. Dietz, K., Heesterbeek, J.: Daniel Bernoulli's epidemiological model revisited. Math. Biosci. **180**, 1–21 (2002)
8. ETHW: Engineering and technology history wiki. https://ethw.org/Michio_Sugeno
9. Google: Google Scholar – https://scholar.google.com/. Accessed September 2020
10. Jafelice, R.M., Bertone, A.M.A.: Minicurso conjuntos fuzzy do tipo 2 intervalar: Teoria e aplicações. In: Minicurso (ed.) IV Congresso Brasileiro de Sistemas Fuzzy, pp. 1–44. Campinas (2016, in Portuguese)
11. Jafelice, R.M., Lodwick, W.A.: Interval analysis of the HIV dynamics model solution using type-2 fuzzy sets. Math. Comput. Simul. **180**, 306–327 (2021)
12. Jafelice, R.M., Silva, P.N.: Studies on population dynamics using cellular automata. In: Salcido, A. (ed.) Cellular Automata: Simplicity Behind Complexity, pp. 105–130. Intech, London (2011)
13. Jafelice, R., Barros, L., Bassanezi, R., Gomide, F.: Fuzzy modeling in asymptomatic HIV virus infected population. Bull. Math. Biol. **66**, 1463–1942 (2004)
14. Jafelice, R.M., Bechara, B.F.Z., Barros, L.C., Bassanezi, R.C., Gomide, F.: Cellular automata with fuzzy parameters in microscopic study of positive HIV individuals. Math. Comput. Model. **50**, 32–44 (2009)
15. Jafelice, R.S.M., Almeida, C.G., Meyer, J.F.C.A., Vasconcelos, H.L.: Fuzzy parameters in a partial differential equation model for population dispersal of leaf-cutting ants. Nonlinear Anal. Real World Appl. **12**, 3397–3412 (2011)
16. Jafelice, R.S.M., Pereira, B.L., Bertone, A.M.A., Barros, L.C.: An epidemiological model for HIV infection in a population using type-2 fuzzy sets and cellular automaton. Comput. Appl. Math. **38**(141) (2019)
17. Jang, J.S.R.: Fuzzy modeling using generalized neural networks and Kalman filter algorithm. In: Proceedings of the 9th National Conference on Artificial Intelligence, pp. 762–767 (1991)
18. Jang, J.S.R.: ANFIS: adaptive-network-based fuzzy inference system. IEEE Trans. Syst. Man Cybern. **23**(3), 665–685 (1993)
19. Karnik, N.N., Mendel, J.M.: Introduction to type-2 fuzzy logic systems. In: 1998 IEEE International Conference on Fuzzy Systems Proceedings. IEEE World Congress on Computational Intelligence, vol. 2, pp. 915–920 (1998)
20. Karnik, N.N., Mendel, J.M.: Centroid of a type-2 fuzzy set. Inf. Sci. **132**, 195–220 (2001)
21. Kermack, W.O., McKendrick, A.G.: Contributions to the mathematical theory of epidemics. R. Stat. Soc. **115**, 700–721 (1927)
22. Lotka, A.J.: Elements of Physical Biology. Williams and Wilkins, Philadelphia (1925)
23. Malthus, T.R.: An Essay on the Principle of Population. J. Johnson, London (1798)
24. Mamdani, E.H., Assilian, S.: An experiment in linguistic synthesis with a fuzzy logic controller. Int. J. Man Mach. Stud. **7**, 1–13 (1975)
25. Massad, E., Burrattini, M.N., Ortega, N.R.S.: Fuzzy logic and measles vaccination: designing a control strategy. Int. J. Epidemiol. **28**(3), 550–557 (1999)
26. Massad, E., Ortega, N.R.S., Barros, L.C., Struchiner, C.J.: Fuzzy Logic in Action: Applications in Epidemiology. Studies in Fuzziness and Soft Computing, vol. 232. Springer, Berlin (2008)
27. Mcneill, D., Freiberger, P.: Fuzzy Logic: The Revolutionary Computer Technology That Is Changing Our World. Simon & Schuster, New York City (1994)
28. Mendel, J.M.: Type-2 fuzzy sets and systems: an overview. IEEE Comput. Intell. Mag. **2**(1), 20–29 (2007)
29. Mendel, J.M.: Type-2 fuzzy sets and systems: a retrospective. Informatik-Spektrum **38**(6), 523–532 (2015)
30. Mendel, J.M.: Type-2 fuzzy sets as well as computing with words. IEEE Comput. Intell. Mag. **14**(1), 82–95 (2019)

31. Mendel, J.M., Wu, D.: Perceptual Computing: Aiding People in Making Subjective Judgments. Wiley and IEEE Press, Hoboken (2010)
32. Ortega, N.R.S., Santos, F.S., Zanetta, D.T., Massad, E.: A fuzzy reed–frost model for epidemic spreading. Bull. Math. Biol. **70**, 1925–1936 (2008)
33. Sanchez, E.: Solutions in composite fuzzy relation equations: application to medical diagnosis in Brouwerian logic. In: Fuzzy Automata and Decision Processes, pp. 221–234. M. M. Gupta, North-Holland, Amsterdam (1977)
34. Sanchez, E., Bartolin, R.: Fuzzy inference and medical diagnosis, a case study. Int. J. Biom. Fuzzy Syst. Ass. **1**, 4–21 (1990)
35. Scott, T., Marketos, P.: On the origin of the Fibonacci sequence. MacTutor History of Mathematics archive, University of St Andrews (2014)
36. Silva, J.D.M.: Análise de estabilidade de sistemas dinâmicos p-fuzzy com aplicações em biomatemática. Ph.D. Thesis, IMECC – UNICAMP, Campinas, Brazil (2005, in Portuguese)
37. Silva, J.D.M., Leite, J., Bassanezi, R.C., Cecconcelo, M.S.: Stationary points-I: One-dimensional p-fuzzy dynamical systems. J. Appl. Math. 495864 (2013)
38. Sugeno, M.: Theory of fuzzy integrals and its applications. Ph.D. Thesis, Tokyo Institute of Technology (1974)
39. Sugeno, M., Kang, G.T.: Structure identification on fuzzy model. Fuzzy Sets Syst. **28**, 329–346 (1988)
40. Takagi, T., Sugeno, M.: Fuzzy identification of systems and its applications to modeling and control. IEEE Trans. Syst. Man Cybern. **15**(1), 116–132 (1985)
41. Verhulst, P.F.: Notice sur la loi que la population suit dans son accroissement. Correspondance mathématique et physique **10**, 113–121 (1838)
42. Zadeh, L.: Fuzzy sets. Inf. Cont. **8**, 338–353 (1965)
43. Zadeh, L.A.: Calculus of fuzzy restrictions. In: Zadeh, L.A., Fu, K.S., Tanaka, K., Shimura, M. (eds.) Fuzzy Sets and Their Applications to Cognitive and Decision Processes, pp. 1–39. Proceedings of the U.S.-Japan Seminar on Fuzzy Sets and Their Applications, held at The University of California, Berkeley. Academic Press Inc., Cambridge (1975)
44. Zadeh, L.A.: The concept of a linguistic variable and its application to approximate reasoning–1. Inf. Sci. **8**, 199–249 (1975)
45. Zadeh, L.A.: The birth and evolution of fuzzy logic. Int. J. Gen. Syst. **17**(2–3), 95–105 (1990)

Chapter 2
A Tour of Type-1 and Interval Type-2 Fuzzy Sets Theory

> *The uncertainty is the imperfection of knowledge about the natural process or natural state. The statistical uncertainty is the randomness or error that comes from different sources as we use it in a statistical methodology [11].*
>
> Castro, Castillo and Martínez (2007)

2.1 Type-1 Fuzzy Sets

Starting this chapter is a succinct review of type-1 fuzzy theory, being Definition 2.1 a description of this fundamental concept.

Definition 2.1 A type-1 fuzzy set A is characterized by its membership function μ_A defined in the universe of discourse X and $\mu_A(x) \in [0, 1]$ [50]. In other words, a type-1 fuzzy set A can be identified by the graph of its membership function:

$$A = \{(x, \mu_A(x)) : x \in X, \mu_A(x) \in [0, 1]\},$$

as shown in Fig. 2.1.

The values $\mu_A(x) = 1$ and $\mu_A(x) = 0$ indicate, respectively, the complete membership and non-membership of the element x to A.

A classic set $B \subset X$ can be interpreted as a type-1 fuzzy set, associating as a membership the characteristic function of B, that is,

$$\mu_B(x) = 1 \text{ if } x \in B \text{ and } \mu_B(x) = 0 \text{ if } x \notin B.$$

A unitary set $B = \{x\}$ along with the characteristic function is a type-1 fuzzy set called a singleton. In addition, in fuzzy language, a classic subset is often called a crisp subset.

Definition 2.2 For the real number $\alpha \in (0, 1]$, the α-level (or α-cut) of type-1 fuzzy set A is defined by the set

R. S. da Motta Jafelice, A. M. A. Bertone, *Biological Models via Interval Type-2 Fuzzy Sets*, SpringerBriefs in Mathematics, https://doi.org/10.1007/978-3-030-64530-4_2

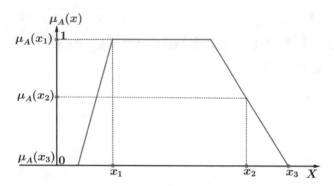

Fig. 2.1 Type-1 fuzzy set

$$[A]^{\alpha} = \{x \in X : \mu_A(x) \geq \alpha\}.$$

Definition 2.3 The support of a type-1 fuzzy set A are all elements of X that have non-zero membership degree in A which is denoted by supp(A), that is

$$\text{supp}(A) = \{x \in X : \mu_A(x) > 0\}.$$

The zero level of the type-1 fuzzy set A is the topological closure of the support of A, that is,

$$[A]^0 = \overline{\text{supp}(A)}.$$

Definition 2.4 A type-1 fuzzy set N is called a type-1 fuzzy number when the universe set X is the set of real numbers $\mathbb{R}$ and all its α-level are closed, limited, and non-empty.

Note that a singleton set and the type-1 fuzzy set given by

$$\mu_A(x) = \{1 \text{ if } x \in [a, b], \ 0 \text{ otherwise}\} \tag{2.1}$$

are type-1 fuzzy numbers. The set A, given by Eq. (2.1) is called a type-1 interval fuzzy number. Figure 2.2 is an exhibit of a type-1 fuzzy set that is not a type-1 fuzzy number. Notice that the α-level at 0.9 is the union of two closed disjoint intervals that is not an interval.

Definition 2.5 Given the type-1 fuzzy sets A and B defined in the universe X, the membership functions representing the fuzzy union, $A \cup B$, intersection, $A \cap B$, and complementary sets, A', are given, respectively, by

$$\mu_{A \cup B}(x) = \max_{x \in X}\{\mu_A(x), \mu_B(x)\}, \tag{2.2}$$

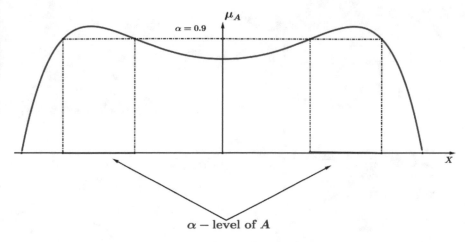

Fig. 2.2 An example of a type-1 fuzzy set that is not a type-1 fuzzy number

$$\mu_{A \cap B}(x) = \min_{x \in X}\{\mu_A(x), \mu_B(x)\},$$

$$\mu_{A'}(x) = 1 - \mu_A(x), \ x \in X.$$

Other important concepts are described in Definition 2.6 and Definition 2.7.

Definition 2.6 A triangular t-norm, $\star$ is an operation $x \star y = r$ where $(x, y) \in [0, 1] \times [0, 1]$, $r \in [0, 1]$ such that satisfies the commutativity, $x \star y = y \star x$; associativity, $(x \star y) \star z = x \star (y \star z)$; monotonicity, which means, if $x \le y$ and $w \le z$ then $x \star w \le y \star z$, and the boundary conditions, $x \star 0 = 0$, $x \star 1 = x$.

Example 2.1 An example of t-norm is $x_1 \star x_2 = \min\{x_1, x_2\}$, for all x_1, $x_2 \in X$ [36] that is denoted by **minimum**, whose name also comes from the fact that it is used to define the intersection of type-1 fuzzy sets [19]. Another t-norm used in the literature is the so-called **product** defined as $x_1 \star x_2 = x_1 \cdot x_2$, for all x_1, $x_2 \in X$, where "·" is the product between real numbers.

Definition 2.7 A s-norm , $\oplus$ is a binary operation $x \oplus y = r$ where $(x, y) \in [0, 1] \times [0, 1]$, $r \in [0, 1]$ satisfies commutativity, that is, $x \oplus y = y \oplus x$; associativity, which means that $(x \star y) \star z = x \star (y \star z)$, monotonicity, that is, if $x \le y$ and $w \le z$ then $x \oplus w \le y \oplus z$, and the boundary conditions, $x \star 0 = x$, $x \star 1 = 1$.

Example 2.2 An example of s-norm is $x_1 \oplus x_2 = \max\{x_1, x_2\}$, for all x_1, $x_2 \in X$ [36] that is called a **maximum** norm, associated with the union operation of type-1 fuzzy sets.

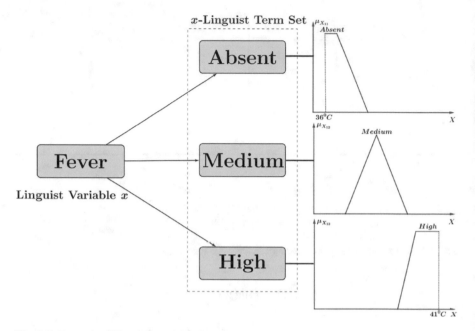

Fig. 2.3 Example of linguistic variable fever

2.1.1   Type-1 FRBS

As Zadeh remarked in his article [51]:

> By a linguistic variable we mean a variable whose values are words or sentences in a natural or artificial language.[...] The main applications of the linguistic approach lie in the realm of humanistic systems-especially in the fields of artificial intelligence, linguistics, human decision processes, pattern recognition, psychology, law, medical diagnosis, information retrieval, economics and related areas.

Example 2.3 An example of a linguistic variable is *fever* as described in Fig. 2.3 [30].

Exercise 2.1 Resolve the following questions:

(a) for the graph of the linguistic variable fever from Fig. 2.3 design an adequate support of the type-1 fuzzy sets of the linguistic terms.
(b) construct type-1 fuzzy sets for the variable fever with the universe being the interval [36,41] that include the linguistic terms to quantify fever as: absent, low, medium, high, and very high.

The linguistic variables are used in the first component of the type-1 FRBS, which is the fuzzification of the variables that make up the system to be modeled. Section 2.1.1.1 explains how to build a type-1 FRBS.

2.1.1.1 Construction of a Type-1 FRBS

Illustrated in Example 2.4 is how to elaborate a type-1 FRBS based on a model from COVID-19. Coronavirus is a type of virus that appeared for the first time on December 26, 2019, in China. On February 11, 2020 the World Health Organization (WHO) [47] announced an official name for the disease that is causing the 2019 novel coronavirus outbreak. The new name of this disease is coronavirus disease 2019, abbreviated as COVID-19. The acronym COVID-19, stands: "CO" for "corona," "VI" for "virus," "D" for disease, and 19 for the year of its discovery. In Sect. 3.6 a mathematical model for the COVID-19 is presented along with a biological explanation of the virus structure.

Example 2.4 A type-1 FRBS is going to be built to quantify the severity of a patient's infection with COVID-19. With the knowledge of an area expert, the main input variables are determined: fever, difficulty breathing, and intensity of dry cough. The structure of the idealized model is shown in Fig. 2.4 which describes the input and output variables.

The components of the type-1 FRBS for its construction are:

- **Fuzzification:** in this component decisions are made to choose the supports, formats, quantities, and categories of the corresponding linguistic terms, of all input and output variables, with the help of a subject matter expert.

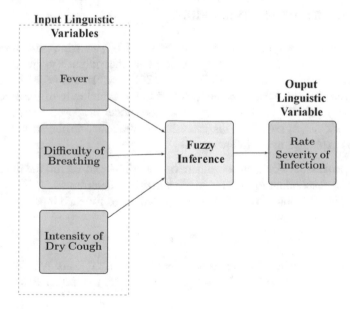

Fig. 2.4 The scheme describes the relationship between linguistic variables

In Example 2.4, the support chosen for the output variable, severity of the COVID-19 patient's infection, is the interval [0, 1]. In the graph of Fig. 2.3 is shown some information related with the fuzzification of the input variable fever.

- **Rule Base:** this component is composed of a collection of propositions in the form If … then … where the antecedent is formed by relations of the linguistic variables and the consequent that can be a fuzzy set or a function. Each of these propositions can, for example, be described input and output variables, with the help of a subject matter expert. In Example 2.4 a rule is:

   If (fever is high) and (difficulty breathing is great) and (intensity of dry cough is high), then (severity of infection is very high).

- **Fuzzy Inference:** in this component each fuzzy proposition is translated mathematically using approximate reasoning techniques [36]. Mathematical operators are selected to define the fuzzy relationship that models the rule base. This way, the fuzzy inference machine is of fundamental importance for the success of the fuzzy system, as it provides the output from each input fuzzy variable and the relationship defined by the rules base. The choice of the type of fuzzy inference method depends on the information provided by the area expert or the literature.

- **Defuzzification:** this component transforms the fuzzy set of an inference output into a real number that represents it. In Example 2.4, after defuzzification, a real vector is obtained in the interval [0, 1] that represents the severity of a patient's infection with the COVID-19.

2.1.1.2 Architecture of a Type-1 FRBS

After having built a type-1 FRBS as in Sect. 2.1.1.1, in which the fuzzy input $(x_1, x_2, \ldots, x_n) \in X_1 \times X_2 \times \ldots \times X_n$ and the output $(y_1, y_2, \ldots, y_p) \in Y_1 \times Y_2 \times \ldots \times Y_p$, wherein X_i $i = 1, 2, \ldots, n$, Y_l, $l = 1, 2, \ldots, p$ are given universes, have been prepared, proceed to the calculation of the crisp output $y' = (y_1(x'), y_2(x'), \ldots, y_p(x')) \in \mathbb{R}^p$ of the model proposed from crisp input $x' = (x_1', x_2', \ldots, x_n') \in \mathbb{R}^n$. This procedure is described in the flow diagram of Fig. 2.5, as explained in the following:

1. **Input processor:** For each coordinate of the crisp input vector, the membership degree corresponding to each linguistic variable is determined.
2. **Rule Base:** Denoting [32] the set of linguistic terms for each input x_i

$$T_{x_i} = \left\{ X_{ij}, j \in I_i \right\}, \quad i = 1, 2, \ldots, n, \tag{2.3}$$

where $I_i \subset \mathbb{N}$ is a set of indexes ordered from 1, X_{ij}, are type-1 fuzzy sets defined in the universes X_i. The output, $(y_1, y_2, \ldots, y_p)$ has an associated set

$$T_{y_s} = \left\{ Y_s^k, k \in I_s \right\}, \quad s = 1, 2, \ldots, p, \tag{2.4}$$

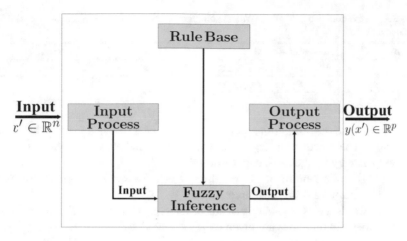

Fig. 2.5 Architecture of the type-1 fuzzy rule-based system

where $I_s \subset \mathbb{N}$ is a set of indexes ordered from 1, and Y_s^k is a type-1 fuzzy set, defined in the Y_s universe, for every $k \in I_s$ or as a function of the input variables. From now on and for sake of simplicity, when there is only one input it is denoted as x. Similarly, when there is only one output it is denoted as y. Furthermore, T_{y_1} is denote by T_y and Y_1^k by Y^k.

The rule $R^m, m \in \mathbb{N}$, $m = 1, 2, \ldots,$ M of type-1 FRBS are defined by

$$R^m : \text{If } x_1 \text{ is } F_1^m \text{ and } x_2 \text{ is } F_2^m \text{ and } \ldots \text{ and } x_n \text{ is } F_n^m \text{ then } y \text{ is } G^m, \qquad (2.5)$$

where F_k^m, $k = 1, \ldots, n$ is an element of set T_{x_i} defined in Eq. (2.3) and G^m is an element of set T_{y_s} defined in Eq. (2.4).

3. **Fuzzy Inference:** The inference block performs the logical operations for type-1 fuzzy sets based on fuzzy rules. Presented are two types of fuzzy inference method: Mamdani [28] and Takagi–Sugeno–Kang [43, 44]. The basic differences between these methods lie in the type of consequences and the process of defuzzification. For simplicity, rule models with two inputs and one output are illustrated.

- **Method of Mamdani**

  Mamdani inference method aggregates the rules through the logical operator $s-$norm, $\oplus$, as **maximum** (see Example 2.2) and, in each rule, the logical operator $t-$norm, $\star$, considered to be the **minimum** (see Example 2.1). Given the following rules:

$$R^1 : \text{If } x_1 \text{ is } X_{11} \text{ and } x_2 \text{ is } X_{21} \text{ then } y \text{ is } Y^1;$$
$$R^2 : \text{If } x_1 \text{ is } X_{12} \text{ and } x_2 \text{ is } X_{22} \text{ then } y \text{ is } Y^2.$$

Table 2.1 Activation degrees of the rules and their corresponding consequent for Mamdani inference method

Rule n⁰	Activation degree	⇒	Consequent
R^1	$f^1 = \mu_{X_{11}}(x_1') \star \mu_{X_{21}}(x_2')$	⇒	Y^1
R^2	$f^2 = \mu_{X_{12}}(x_1') \star \mu_{X_{22}}(x_2')$	⇒	Y^2

The degree of truth, or degree of activation, of a rule is a value y between 0 and 1. These activation degrees of the rules R^1 and R^2 are shown in Table 2.1 associated with their respective consequent.

Note that the t-norm of the **product** is also known in the literature as the Larsen norm [24], see Example 2.1.

Afterwards, being $y \in Y$ it is obtained the type-1 fuzzy sets B^1 and B^2 which membership functions are given by

$$\mu_{B^1}(y) = f^1 \star \mu_{Y^1}(y) \text{ and } \mu_{B^2}(y) = f^2 \star \mu_{Y^2}(y).$$

Finally, the type-1 fuzzy set, B, results from the inference, using the s-norm $\oplus$ as being the **maximum** is given by the following membership function:

$$\mu_B(y) = \mu_{B^1}(y) \oplus \mu_{B^2}(y). \tag{2.6}$$

Illustrated in Fig. 2.6 is the composition rule known as the **maximum-minimum**.

The output y' is obtained by defuzzification of the output fuzzy set B. In Fig. 2.6 it is shown that $B = B^1 \cup B^2$.

- **Method of Takagi–Sugeno–Kang (TSK)**

  In this case, the consequent of each rule is a function of the input variables. For example, it is assumed that the function that maps the input and output for each rule is a linear combination of the inputs. See the following rules:

  $$\begin{aligned} R^1 &: \text{ If } x_1 \text{ is } X_{11} \text{ and } x_2 \text{ is } X_{21} \text{ then } y \text{ is } Y^1 = p_1 x_1 + q_1 x_2 + r_1; \\ R^2 &: \text{ If } x_1 \text{ is } X_{12} \text{ and } x_2 \text{ is } X_{22} \text{ then } y \text{ is } Y^2 = p_2 x_1 + q_2 x_2 + r_2; \end{aligned} \tag{2.7}$$

  where p_1, p_2, q_1, q_2, r_1, r_2 are real numbers. The activation degree of each rule R^1 and R^2 associated with its respective consequent is shown in Table 2.2. The t-norms, $\star$, the most used are the **minimum** and the **product** (see Example 2.1).

  Figure 2.7 illustrates the flow diagram of the TSK inference method for two inputs and one output.

4. **Defuzzification**: this component determines the real output of the modeled system from the inference method. Depending on this choice the following alternatives are available:

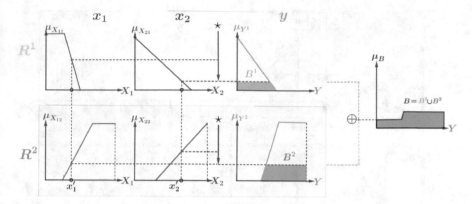

Fig. 2.6 Mamdani inference method with composition **maximum-minimum**

Table 2.2 Activation degree of the rules and their corresponding consequent for TSK inference method

Rule $n^{\underline{o}}$	Activation degree	$\Rightarrow$	Consequent
R^1	$f^1 = \mu_{X_{11}}(x_1') \star \mu_{X_{21}}(x_2')$	$\Rightarrow$	Y^1
R^2	$f^2 = \mu_{X_{12}}(x_1') \star \mu_{X_{22}}(x_2')$	$\Rightarrow$	Y^2

- **Defuzzification in the Mamdani Inference Method:** there are several methods, among which, the most used is that of Center of Gravity or Centroid. To calculate it for the fuzzy set, B, obtained in Eq. (2.6), an average like the weighted average for data distribution is performed, with the difference that the weights are the values of μ_B. For a discrete domain this value given by

$$y' = \frac{\sum_{i=0}^{n} z_i \mu_B(z_i)}{\sum_{i=0}^{n} \mu_B(z_i)}, \quad z_i \text{ in the domain of } \mu_B. \tag{2.8}$$

- **Defuzzification in the TSK Inference Method:** the output y of a system of the TSK inference method is generated from the real inputs obtained by the weighted average of the outputs of each rule, using the degree of activation of these rules as the weight. In fact, using the rule base of Eq. (2.7), it is obtained

$$y' = \frac{f^1 Y^1 + f^2 Y^2}{f^1 + f^2}. \tag{2.9}$$

In the case that T_y is formed by singleton sets, the Mamdani and TSK inference methods produce the same output values, because the defuzzification in the Mamdani inference method, by the center of gravity, is equal to the weighted average in the TSK inference method.

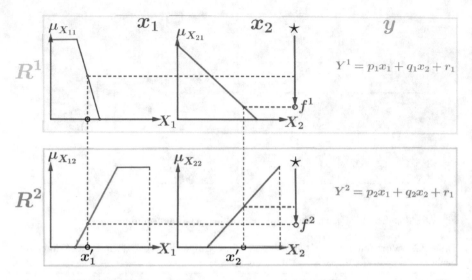

Fig. 2.7 TSK fuzzy inference system with two inputs and one output

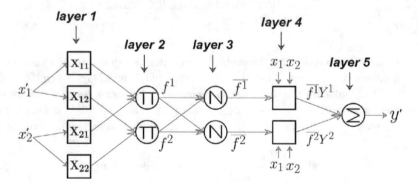

Fig. 2.8 ANFIS structure. Image build based on the scheme extracted from Jang [21]

2.1.2   *Adaptive Neuro-Fuzzy Inference Systems (ANFIS)*

Proposed by Jang [21], ANFIS is a training routine for TSK inference method, which, based on a set of data, uses a learning algorithm to identify the system parameters based on fuzzy rules [45].

To facilitate the understanding of the structure of the ANFIS neural network, a system is presented that has two inputs x_1 and x_2 and an output y [21]. A base with two rules is given by Eq. (2.7).

Note that the inference system shown in Fig. 2.7 corresponds to the ANFIS architecture presented in Fig. 2.8.

Thus, the structure of ANFIS by layers is given by:

Layer 1: reads the input crisp values and determines its membership degree according to the equation

$$O_i^1 = \mu_{X_{1i}}(x_1'), \ i = 1, 2 \ \text{ and } \ O_i^1 = \mu_{X_{2i-2}}(x_2'), \ i = 3, 4.$$

The parameters of this layer are called premise parameters and are modified by the backpropagation algorithm [41] and gradient descent for the network learning.

Layer 2: applies the $t-$norm, $\star$ (**minimum** or **product**, see Example 2.1) represented in the node Π of Fig. 2.8, among the values of the input membership functions, a result that is known as firing strength, as shown in Table 2.2.

Layer 3: is the normalization shown in node N of Fig. 2.8, the activation degree of each rule [21]. In fact, outputs of layer 2 are divided by the sum of all outputs, normalizing the activation degrees for each rule. In short, it is a normalization of firing strength, known as normalized firing strength and calculated as

$$\overline{f^i} = \frac{f^i}{f^1 + f^2}, \ i = 1, 2.$$

Layer 4: is one of the adaptive layers, with the input of layer 3 as the consequent parameters calculated as

$$O_i^4 = \overline{f^i} Y^i = f^i(p_i x_1' + q_i x_2' + r_i),$$

where $(p_i, q_i, r_i), i = 1, 2$, is the set of consequent parameters, which is modified by an adaptation of the network. The values returned by this layer are the defuzzified ones and these values are passed to the last layer to return the final output.

Layer 5: is a fixed layer corresponding to the defuzzification method, consisting of a single node, $\sum$, shown in Fig. 2.8. In this layer it is calculated:

$$O_i^5 = \sum_{i=1}^{2} \overline{f^i} Y^i,$$

that is, the output y' is obtained by a weighted average value.

Finally, the learning of the ANFIS neural network takes place in two steps:

First Step: forward pass of the hybrid learning algorithm, functional signals go forward up to layer 4. The consequent parameters are determined by the least square estimate.

Second Step: the backward pass in which the error rates propagate backward and the premise parameters are updated by the gradient descent.

As a final remark and a historical note is highlighted that the least-squares method is usually credited to Carl Friedrich Gauss (1795) [42] and it was first published by Adrien-Marie Legendre [25]. Gradient descent was originally proposed by Cauchy in 1847 [12] as remarked by Lémarechal [26].

2.1.3 Partially Fuzzy Systems

This technique was introduced by Peixoto [37], Peixoto et al. [38], and Barros et al. [4] to create a model that studies biological phenomena which do not have sufficient information. On the other hand, qualitative information from specialists had allowed rules to be proposed that relate (at least partially), to the state variables, with their own variations.

Differential equations are tools for modeling phenomena whose state variables are related to their temporal variables. This relationship is established from parameters or functions that need to be measured or estimated. However in most cases, it is necessary to collect a large number of data to describe the phenomenon to be modeled well. In addition, in many of these phenomena the relationship between variables and their variations is imprecise. This characteristic makes it difficult to model the phenomenon using differential equations since they depend on the precision of the parameters used. For theoretical fuzzy approaches of the differentiation procedure see [16].

Proposed in this section is the adoption of FRBS incorporate inaccurate information on the variables, on the variations, and on their relations with the variables, thus allowing the mathematical treatment of these uncertainties. Such systems are called partially fuzzy or for short, p-fuzzy systems. The term comes from the fact that in ordinary differential equations the systems are partially fuzzy in the sense that the direction field of the Initial Value Problem, IVP, in question is partially known [3, 14, 15]. However, its solution is real and at each time t, a corresponding value is obtained after a defuzzification process. In mathematical terms one can define that IVP given by $\dfrac{dx}{dt} = f(x(t))$, $x(t_0) = x_0$ is p-fuzzy if f is partially known and can be described through a fuzzy rule base.

Definition 2.8 It is denominated by a type-1 p-fuzzy system when this system is constructed through type-1 FRBS and is an interval type-2 p-fuzzy system when it is built with an interval type-2 FRBS.

To obtain the trajectory of a p-fuzzy system as a function of time, a type-1 FRBS is needed, in which the input variable is the state variable and the output variable is the variation rate. In the applications shown in Chap. 3, the type-1 FRBS has been carried out through two decisions:

Decision 1:   information from the literature, knowing the experimental data or data collection, using the TSK inference method, obtained through ANFIS;

Fig. 2.9 Scheme to obtain the trajectory of the p-fuzzy system

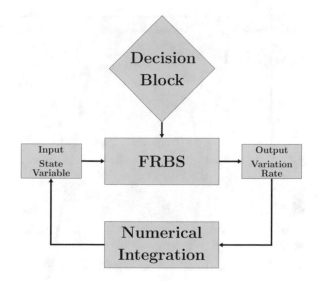

Decision 2: information from the literature and/or knowledge from a subject matter expert, using Mamdani inference method.

Shown in Fig. 2.9 is the scheme to determine the trajectory of the p-fuzzy system. In the decision block, one of the two alternatives is used. The process begins with the initial condition of the input block, following by the flowchart to obtain the variation rate output. Note that a numerical integration method is required to determine the input variable for the next iteration. When one knows the differential equation for the model, it can compare the solution of this equation with the trajectory obtained by the p-fuzzy system.

In order to better understand the exposed methodology, the following is an example of the effectiveness and versatility of p-fuzzy systems in modeling real situations.

Example 2.5 Verhulst's model [46] assumes that a population, living in a given environment, must grow to a sustainable limit, that is, it tends to stabilize. This model is essentially the modified Malthus model [27], considering the growth rate as being proportional to the population at each instant. Therefore, extracting information from the literature, consider a determined population with nonconstant intrinsic growth rate which is a linear function, β, depending on the population, P, at each instant, as follows:

$$\beta(P) = r\left(1 - \frac{P}{P_\infty}\right), \qquad r = 0.038, \qquad P_\infty = 282,597,030. \qquad (2.10)$$

Notice that the calculation of r and P_∞, the latter being the constant that represents the stabilization of the sustainable limit for the population, is done from the data information of the population and a methodology shown in [5].

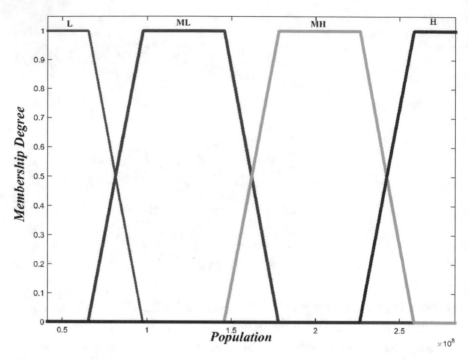

Fig. 2.10 Membership functions for the input variable P

Based on this information and **Decision 1**, consider the initial population of $P_0 = 41,236,315$ individuals, total population when it reaches instant 150, and a type-1 FRBS whose output is the population variation rate. To proceed with the modeling, an ANFIS training is used in the interval $D = [P_0, \ P_\infty]$ as the input variable domain. Furthermore, if $P \in D$, then the output variable is $r \left(1 - \frac{P}{P_\infty}\right) P$, where both, r and P_∞, are defined in Eq. (2.10).

The system is constructed with the linguistic input variable, population P, and the output being the variation rate, $\frac{dP}{dt}$. For the population P, the linguistic terms are Low (L), Medium Low (ML), Medium High (MH), and High (H), represented in Fig. 2.10 by trapezoidal membership functions.

The output variable is a polynomial function $f(P) = mP + n$, where the parameters m and n are determined by ANFIS. The rule base is built from the variation rate as a function of input P composed of

- If P is L then $\dfrac{dP}{dt}$ is $Y^1 = -0.0092P + 2,649,472$;
- If P is ML then $\dfrac{dP}{dt}$ is $Y^2 = 0.0018P + 2,381,125$;
- If P is MH then $\dfrac{dP}{dt}$ is $Y^3 = -0.0151P + 5,191,076$;

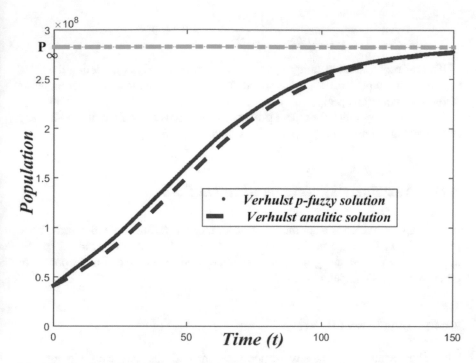

Fig. 2.11 Graphs of the trajectory's type-1 p-fuzzy system and the analytic solution

- If P is H then $\dfrac{dP}{dt}$ is $Y^4 = -0.0287P + 8{,}198{,}277$.

Following the flowchart of Fig. 2.9, the first output from the initial population P_0 is calculated, proceeding with the numerical integration method of Simpson's Rule [7] to obtain the next value $P_f(t)$ of the population, up to $t = 150$.

The IVP associated to this model is given by

$$\frac{dP}{dt} = \beta(P)P, \quad P(0) = P_0,$$

where β is defined in Eq. (2.10) and its solution is

$$P(t) = \frac{P_\infty P_0}{(P_\infty - P_0)e^{-rt} + P_0}. \tag{2.11}$$

Shown in Fig. 2.11 is the analytic solution, $P(t)$ given by Eq. (2.11) and the trajectory of the p-fuzzy system $P_f(t)$ at time $t \in [0, 150]$. Notice that $\beta(P)$ approximates to zero when the solution is close to P_∞. Comparing both solutions through the accuracy measure, maximum of relative error at each instant, results in

$$\max_{t \in [0,150]} \frac{|P(t) - P_f(t)|}{|P(t)|} = 0.1169.$$

The p-fuzzy system technique presents a great potential for modeling dynamic systems using type-1 FRBS, more specifically, the potential to model autonomous differential equations [19].

The technique applied for type-1 p-fuzzy systems is used for interval type-2 fuzzy sets in biological phenomena applications in the Chap. 3.

2.1.4 New Fuzzy System Identification Data

A new method, which is a combination of a fuzzy clustering procedure and a TSK [43, 44] fuzzy inference method, has been developed by Martins et al. [29]. The approach has two advantages: gain in accuracy and greater speed of response, in other words, less computational effort.

2.1.4.1 Data Fuzzy Clustering

Clustering techniques are mostly unsupervised methods that can be used to organize groups based on a mathematical similarity criterion. The clustering algorithms can show underlying structures, and their applications are several, such as: image processing, pattern recognition, systems identification, among others. The difference in fuzzy clustering is the way the data are partitioned: each element has a certain membership degree in relation to each cluster, that is, values between 0 and 1.

Given $Z = \{z_1, z_2, \ldots, z_N\}$ a subset of data belonging to the universe X of type-1 fuzzy sets $C_1, C_2, \ldots, C_c$, called clusters of fuzzy clustering, it is denoted by μ_{ij}, $1 \leq i \leq N$ and $1 \leq j \leq c$, the membership degree of the element z_i to the cluster C_j. Thus, the characteristics of fuzzy clustering are:

1. Each z_i element has a membership degree $0 \leq \mu_{ij} \leq 1$ to each cluster C_j;
2. The sum of all the membership degrees of an element in relation to each cluster must be equal to 1, that is, $\sum_{j=1}^{c} \mu_{ij} = 1$ for each $i = 1, \ldots, N$;
3. The sum of the membership degrees of all the z_i of a cluster C_j cannot be null, that is, $\sum_{i=1}^{N} \mu_{ij} > 0$, $j = 1, \ldots, c$.

In this case, being $n \in \mathbb{N}$, the i row of the data set Z is

$$z_i = (x_{i1}, x_{i2}, \ldots, z_{in}), \quad i = 1, \ldots, N;$$

where z_{in} represents the data output that depends on the inputs $x_{i1}, x_{i2}, \ldots, x_{i(n-1)}$. Associated with Z, the so-called membership matrix is given by

$$
U = \begin{bmatrix}
\mu_{11} & \cdots & \mu_{1j} & \cdots & \mu_{1c} \\
\mu_{21} & \cdots & \mu_{2j} & \cdots & \mu_{2c} \\
\vdots & \vdots & \ddots & & \vdots \\
\mu_{N1} & \cdots & \mu_{Nj} & \cdots & \mu_{Nc},
\end{bmatrix},
\tag{2.12}
$$

where μ_{ij} is the membership degree of the point z_i to the cluster C_j.

A widely used form of similarity, which meets geometric objectives, is the distance defined by

$$
d^2_{ij,M_j}(z_i, v_j) = (z_i - v_j)^T M_j (z_i - v_j), \quad 1 \le j \le c, \quad 1 \le i \le N, \tag{2.13}
$$

where M_j is a definite positive symmetric matrix and $v_j \in \mathbb{R}^n$ are called the cluster centers C_j. The shorter the distance from the element to the cluster center, the membership degree is greater, being 1 when both coincide.

One of the first fuzzy clustering methods that became known was Fuzzy C-means, developed by James Dunn [13] and enhanced by James Bezdek [6]. After the emergence of the fuzzy C-means algorithm, a large family of clustering algorithms appeared that employed new means of carrying out the clustering process. The Gustafson-Kessel [17] algorithm is part of the set of improvements made from the fuzzy C-means. This algorithm is a powerful technique with a wide range of applications. Its main feature is the geometric adaptation of the cluster format estimated from the M matrix which is based on the Mahalanobis distance [22]. Thus, at each iteration, k, the geometric structure of the clusters change, while the C-means algorithm uses the same distance in all iterations.

In this study, the centers of the clusters are chosen randomly from the data z_i, and the matrix $M_j = I$ of Eq. (2.13), I being the identity matrix. Next, a relationship is sought between the Mahalanobis distance and the membership degree of the sampled points, thus formulating an optimization problem

$$
\text{Minimum} \sum_{j=1}^{c} \sum_{i=1}^{N} \mu_{ij}^2 d^2_{ij,M_j}, \tag{2.14}
$$

for some symmetrical positive defined matrix M_j and restricted to conditions determined by the fuzzy clustering characteristics, as being: $0 \le \mu_{ij} \le 1$, $i = 1, \ldots, N$, $j = 1, \ldots, c$, and

$$
\sum_{i=1}^{N} \mu_{ij} > 0, \ j = 1, \ldots, c; \quad \sum_{j=1}^{c} \mu_{ij} = 1, \ i = 1, \ldots, N. \tag{2.15}
$$

Solving the problem (2.14) and (2.15) through the Lagrange multiplier method it is obtained

$$\mu_{ij} = \frac{1}{\sum_{k=1}^{c} \left(\frac{d_{ij,M_j}^2}{d_{ik,M_j}^2} \right)^{\frac{1}{2}}}, \quad i = 1, 2, \ldots, N, \quad j = 1, 2, \ldots, c. \tag{2.16}$$

Observe that a relationship between the distance d_{ij,M_j} and the membership degree is determined. The matrix U is built using Eq. (2.16) in the different iterations.

Furthermore, the new cluster centers are calculated, as being $v_j = \dfrac{\sum_{i=1}^{N} \mu_{ij}^m z_i}{\sum_{i=1}^{N} \mu_{ij}^2}$ [2].

In the next step, the so-called covariance matrix is calculated:

$$F_j = \sum_{i=1}^{N} \frac{\mu_{ij}^2}{\sum_{i=1}^{N} \mu_{ij}^2} (z_i - v_j)(z_i - v_j)^T, \quad j = 1, 2, \ldots, c,$$

which allows an update of the matrix M_j through the formula

$$M_j = \det(F_j)^{\frac{1}{n}} F_j^{-1}, \quad j = 1, 2, \ldots, c,$$

F_j^{-1} being the inverse matrix of F_j, thus defining a new distance for each iteration.

Therefore, in each iteration is obtained a new matrix $U^k = (u_{ij}^k)$ containing the membership degrees of each point to the clusters. Finally, the error value in the iteration is checked, which is calculated from the norm of the maximum difference between the membership matrices before and after the alteration of M_j, that is

$$\text{error} = \text{maximun}_{\substack{i \in [1, N], \\ j \in [1, c]}} \left| u_{ij}^k - u_{ij}^{k+1} \right|.$$

The stop criterion for the algorithm is when error $\leq \varepsilon$, being ε a predetermined number.

2.1.4.2   A New Takagi–Sugeno–Kang Inference

As a consequence of clustering procedure, four fundamental elements are obtained for the construction of TSK inference method:

1. the centers of clusters v_j which are the centers of Gaussian membership functions, corresponding to the antecedents of the inference;
2. the membership matrix U;

3. the projection of α-levels in the first axis x_1 corresponding to the antecedents, A_j, given by

$$[A_j]^\alpha = \{x_{i1} \in X, \mu_{ij}(z_i) \geq \alpha\}, \quad j = 1, \ldots, c, \tag{2.17}$$

where $\alpha \in [0.5, \ 1]$ is determined by an optimization process in order to approximate the points of the projected cluster C_j to a Gaussian function;

4. the standard deviation of the Gaussian function is given by

$$\sigma_j = \beta(\max A_j - \min A_j), \tag{2.18}$$

where the parameter β is determined by the same optimization process as the parameter α.

Thus, the antecedents' membership functions of the inference are defined by

$$\varrho_j(x) = \exp\left(-\frac{(x - v_j)^2}{2\sigma_j^2}\right), \quad j = 1, 2, \ldots, c. \tag{2.19}$$

The local linear regression of the grouped elements is then performed using a multiple linear regression [1] in which the attributes of the output data are incorporated. Then, the vector $\theta_j = (\theta_{j0}, \theta_{j1}, \ldots, \theta_{jn})$ is obtained to establish the fuzzy rules, equal to the number of clusters. In fact, given any input vector $(x_1, x_2, \ldots, x_{n-1})$, it is defined

$$\text{Rule}^j : \text{ If } x_1 \text{ is } A_j \text{ then } Y^j(x_1) = \theta_{j0} + \sum_{k=1}^{n-1} \theta_{jk} x_k, \quad j = 1, \ldots, c,$$

Finally, the defuzzification is done, in the classic way, calculating the weighted average as follows:

$$y' = \frac{\sum\limits_{j=1}^{c} \varrho_j \cdot Y^j}{\sum\limits_{j=1}^{c} \varrho_j}. \tag{2.20}$$

The steps of the methodology can be summarized in the flowchart, shown in Fig. 2.12. This fuzzy system identification is used in an application of Sect. 3.6.

Fig. 2.12 Flowchart of the
fuzzy system identification

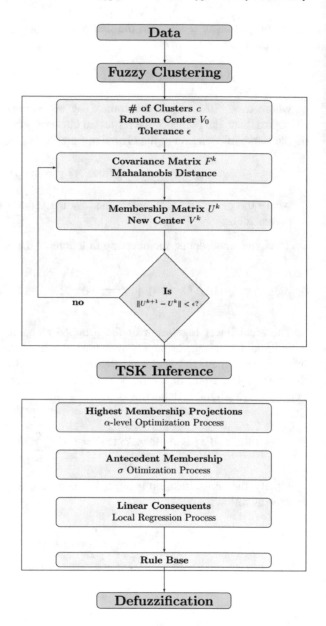

2.2   Type-2 Fuzzy Sets

In Mendel et al. [35], the authors rewrite and explain many definitions and notations
of type-2 fuzzy sets. The following definitions are based on that article.

Definition 2.9 Given X, a universe of discourse, a type-2 fuzzy set $\tilde{A}$ in X is the graph of the function

$$\mu_{\tilde{A}} : X \times [0, 1] \Rightarrow \quad [0, 1]$$
$$(x, u) \quad \Rightarrow \mu_{\tilde{A}}(x, u),$$

where $\mu_{\tilde{A}}$ is called the membership function of $\tilde{A}$. In other words, the type-2 fuzzy set of X, $\tilde{A}$, is given by

$$\tilde{A} = \left\{ ((x, u), \mu_{\tilde{A}}(x, u)) | (x, u) \in X \times [0, 1], \mu_{\tilde{A}}(x, u) \in [0, 1] \right\}.$$

Shown in Fig. 2.13 is an example of a discrete type-2 fuzzy set $\widetilde{A}$ given by

$$\widetilde{A} = \{((-2, 0.1), 0.4), ((-2, 0.5), 0.8), ((-2, 0.9), 1),$$
$$((2.5, 0.3), 0.8), ((2.5, 0.5), 1), ((2.5, 0.7), 0.4), ((2.5, 1), 0.2),$$
$$((4.5, 0.1), 0.2), ((4.5, 0.4), 1), ((4.5, 0.9), 0.7)\}.$$

Definition 2.10 The secondary membership function, $\mu_{\widetilde{A}(x)}(u)$ of x, is the membership function of the type-1 fuzzy set, $\widetilde{A}(x)$, which is obtained by cutting a plane parallel to the u axis by x. The support of $\widetilde{A}(x)$ (see Definition 2.3) is denoted by I_x, that is

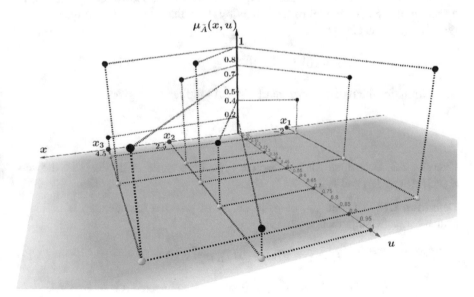

Fig. 2.13 The type-2 fuzzy set, $\widetilde{A}$. The x-coordinate of the points correspond to $\widetilde{A}$ are: $x_1 = -2$, $x_2 = 2.5$, and $x_3 = 4.5$

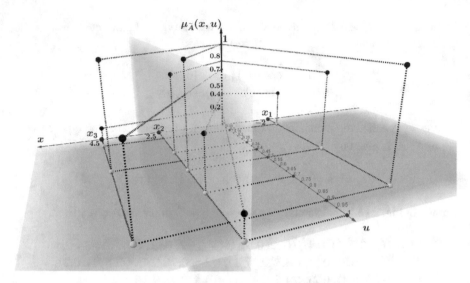

Fig. 2.14 The discrete type-2 fuzzy set $\widetilde{A}$ shown in Fig. 2.13 and I_{x_2}

$$I_x = \{u|u \in [0, 1], \mu_{\widetilde{A}}(x, u) > 0\}. \tag{2.21}$$

Note that $\mu_{\widetilde{A}(x)}(u)$ is a membership function because $\mu_{\widetilde{A}(x)}(u) \in [0, 1]$.

Shown in Fig. 2.14 is the discrete type-2 fuzzy set, $\widetilde{A}$, shown in Fig. 2.13, with a section of a plane parallel to the u axis by the point $x_2 = 2.5$. The fuzzy set of type-1, $\widetilde{A}(2.5)$, is given by

$$\widetilde{A}(2.5) = \{(0.3, 0.8); (0.5, 1); (0.7, 0.4); (1, 0.2)\}.$$

In addition, for the type-2 fuzzy set of Fig. 2.14 the set I_{x_2} is given by

$$I_{x_2} = \{0.3, 0.5, 0.7, 1\}.$$

Exercise 2.2 Determine for the discrete type-2 fuzzy set $\widetilde{A}$ of Fig. 2.14 the sets $\widetilde{A}(x_1)$, $\widetilde{A}(x_3)$, I_{x_1}, and I_{x_3}.

Definition 2.11 The set, J_x, of pairs (x, u) of positive membership is called x primary membership, that is,

$$J_x = \{(x, u)|u \in [0, 1], \mu_{\widetilde{A}}(x, u) > 0\}. \tag{2.22}$$

Note that $J_x = \{x\} \times I_x$. As $I_x \subset \mathbb{R}$ then if it is connected it is an interval.

Example 2.6 From Eq. (2.22), the set J_{x_2} of the discrete type-2 fuzzy set $\widetilde{A}$, shown in Fig. 2.14, is given by

$$J_{x_2} = \{(2.5, 0.3), (2.5, 0.5), (2.5, 0.7), (2.5, 1)\}.$$

Exercise 2.3 Determine for the discrete type-2 fuzzy set $\widetilde{A}$ of the Fig. 2.14, the sets J_{x_1} and J_{x_3} defined in Eq. (2.22).

Definition 2.12 Domain of Uncertainty of $\widetilde{A}$, $DOU(\widetilde{A})$ is the support of $\widetilde{A}$, that is,

$$DOU(\widetilde{A}) = \{(x, u) \in X \times [0, 1] | \mu_{\widetilde{A}}(x, u) > 0\}.$$

The $DOU(\widetilde{A})$ is the set of points highlighted in Fig. 2.13 with yellow color in the plane xOu. Note that, in general, $DOU(\widetilde{A}) = \bigcup_{x \in X} J_x$.

Definition 2.13 The upper and lower membership functions, denoted by $\overline{\mu}_{\widetilde{A}}(x)$, $\underline{\mu}_{\widetilde{A}}(x)$, $x \in X$, are, respectively, defined by

$$\overline{\mu}_{\widetilde{A}}(x) = \sup\{u | u \in [0, 1], \mu_{\widetilde{A}}(x, u) > 0\} = \sup I_x, \tag{2.23}$$

$$\underline{\mu}_{\widetilde{A}}(x) = \inf\{u | u \in [0, 1], \mu_{\widetilde{A}}(x, u) > 0\} = \inf I_x, \tag{2.24}$$

where $\sup I_x$ is the supremum of the set I_x and $\inf I_x$ is the infimum of the set I_x.

Example 2.7 In the discrete type-2 fuzzy set $\widetilde{A}$, Fig. 2.14, $\overline{\mu}_{\widetilde{B}}(x_2) = 1$ and $\underline{\mu}_{\widetilde{B}}(x_2) = 0.3$.

Exercise 2.4 Determine for the discrete type-2 fuzzy set $\widetilde{A}$ (see Fig. 2.14), $\overline{\mu}_{\widetilde{A}}(x_1)$, $\underline{\mu}_{\widetilde{A}}(x_1)$, $\overline{\mu}_{\widetilde{A}}(x_3)$, and $\underline{\mu}_{\widetilde{A}}(x_3)$.

Definition 2.14 Footprint of Uncertainty of $\widetilde{A}$, $FOU(\widetilde{A})$ is the set

$$FOU(\widetilde{A}) = \{(x, u) | x \in X \text{ and } u \in [\underline{\mu}_{\widetilde{A}}(x), \overline{\mu}_{\widetilde{A}}(x)]\}, \tag{2.25}$$

where $\overline{\mu}_{\widetilde{A}}(x)$ and $\underline{\mu}_{\widetilde{A}}(x)$ are defined in (2.23) and (2.24), respectively.

Observe that if $I_x = [\underline{\mu}_{\widetilde{A}}(x), \overline{\mu}_{\widetilde{A}}(x)]$, then $FOU(\widetilde{A}) = DOU(\widetilde{A})$.

Exhibit in Fig. 2.16 is a continuous type-2 fuzzy set $\widetilde{B}$. The goal to present this specific set is to explore its components, in particular of $FOU(\widetilde{B})$.

Example 2.8 The graph of the surface given by equation $f(x, u) = -\ln(x^2 + u^2)$ is shown in Fig. 2.15. Depicted in Fig. 2.16 is the graph of the type-2 membership function of $\widetilde{B}$, given by

$$z = \mu_{\widetilde{B}}(x, u) = -\ln(x^2 + u^2), \quad 0.36 \leq x^2 + u^2 \leq 1, \quad u \in [0, 1], \quad 0 \leq z \leq 1.$$

Some components of $\widetilde{B}$ to be highlighted are

- The secondary membership function for $x = 0.3$ (see Fig. 2.16) is given by

$$\mu_{\widetilde{B}(0.3)}(u) = -\ln((0.3)^2 + u^2).$$

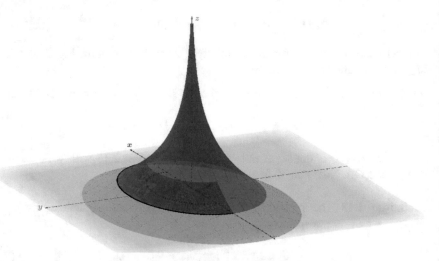

Fig. 2.15 Surface that generates the type-2 fuzzy set $\widetilde{B}$ of Fig. 2.16

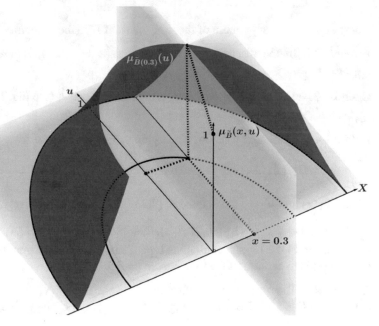

Fig. 2.16 The type-2 fuzzy set $\widetilde{B}$

- Taking $x = 0.3$ then $I_x = [0.51, \ 0.95]$.
- The primary membership J_x corresponding to $x = 0.3$ is the set

$$J_x = \{(0.3, u), u \in I_x\};$$

Fig. 2.17 FOU of the type-2
fuzzy set $\widetilde{B}$

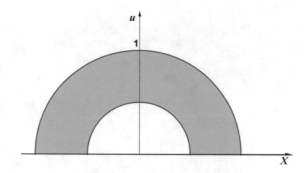

- The $DOU(\widetilde{B})$, that is the $FOU(\widetilde{B})$, because is a continuous type-2 fuzzy set, it is shown in Fig. 2.17.

Exercise 2.5 For the type-2 set $\widetilde{B}$ of Fig. 2.16 analytically write the following elements:

(a) the secondary membership functions for $x = -0.4$ and $x = 0.6$;
(b) I_x and J_x to $x = -0.4$ and $x = 0.6$;
(c) $FOU(\widetilde{B})$.

Definition 2.15 Given the type-2 fuzzy set $\widetilde{C}$, if all values of $\mu_{\widetilde{C}}(x, u)$ are unitary, that is, $\mu_{\widetilde{C}}(x, u) = 1$ for all $(x, u) \in X \times [0, 1]$ is called an interval type-2 fuzzy set.

Example 2.9 The discrete interval type-2 fuzzy set $\widetilde{C}$ described in Fig. 2.18 is determined by a section of $\widetilde{A}$ shown in Fig. 2.13, that is parallel to the xOu plane with height 1. The set $\widetilde{C}$ is given by

$$\widetilde{C} = \{((-2, 0.9), 1), ((2.5, 0.5), 1), ((4.5, 0.4), 1)\}.$$

In Example 2.10 is shown a model described by one continuous interval type-2 fuzzy set. The data of the model has been extracted from the source World Health Organization, WHO [48].

Example 2.10 The research Multicentre Growth Reference Study (MGRS) developed by WHO has provided the data, undertaken between 1997 and 2003 to generate curves for the growth development of infants and young children around the world. The MGRS collected primary growth data and related information from approximately 8,500 children from widely different ethnic backgrounds and cultural settings (Brazil, Ghana, India, Norway, Oman, and the USA). The calculation was performed using the global world references standards for heights of infants, children, and adolescents, given by:

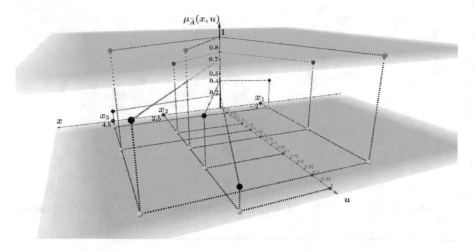

Fig. 2.18 Discrete interval type-2 fuzzy set

1. Median expected height by age (shown as the thick line);
2. Two Standard Deviations, 2SD, above and below the median (shown as the shaded areas in the graph of Fig. 2.19).

The shaded ribbons indicate heights in the range defined as "healthy" growth. Children with height which fall below 2SD are defined as "stunted": having a height too short for their age. Depicted in Fig. 2.19 are the results of the study, where the height values have been normalized: from the 40 to 200 cm scale it has been transformed into the scale from 0.2 to 1. The ranges are determined by the curves $\underline{g} = Median\ Girls - 2SD$ and $\overline{g} = Median\ Girls + 2SD$ where $2SD$ is two standard deviations.

Note that Fig. 2.19 models the following logical statement:

The statement { The height of girls aged between 0 to 19 years is in the range bounded by $\underline{g}$ and $\overline{g}$ } has truth value of 1."

Thus, the set $\widetilde{G}$ is a continuous interval type-2 fuzzy set since $\mu_{\widetilde{G}}(x, u) = 1$ for all $(x, u) \in [0, 19] \times [\underline{g}, \overline{g}]$. In Fig. 2.19 is presented the $FOU(\widetilde{G})$.

In the following, some of the kinds of continuous interval type-2 sets that have been used in the applications of Chap. 3 are depicted. Presented in Fig. 2.20a is the FOU of a triangular interval type-2 fuzzy set with lower membership function that has an image equal to 1; Fig. 2.20b is trapezoidal; Fig. 2.20c is a Gaussian, and Fig. 2.20d is a triangular with lower membership function that does not have an image equal to 1 [11].

As a final remark, note that an interval type-2 fuzzy set, $\widetilde{C}$, is fully represented by its $DOU(\widetilde{C})$, as shown in Fig. 2.21. It should be observed that this figure is the three-dimensional interpretation of the interval type-2 fuzzy set of Fig. 2.20b.

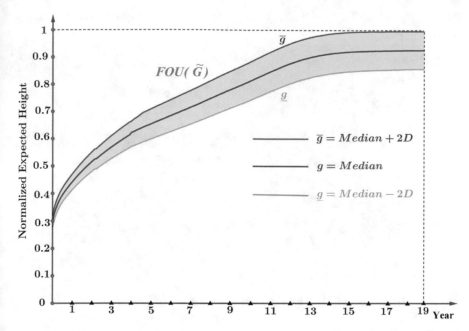

Fig. 2.19 The model of Example 2.10 defines the continuous interval type-2 fuzzy set $\widetilde{G}$ [48]. Graph built with GeoGebra software's tools [18]

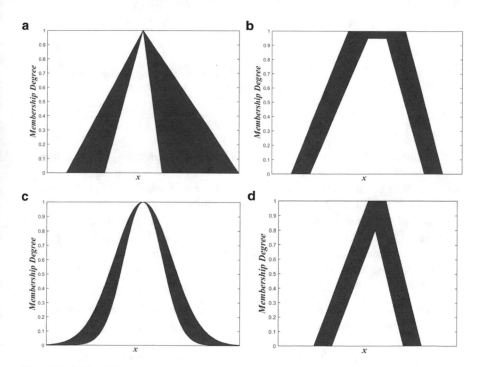

Fig. 2.20 FOUs of interval type-2 fuzzy set examples. (**a**) Triangular. (**b**) Trapezoidal. (**c**) Gaussian. (**d**) Triangular

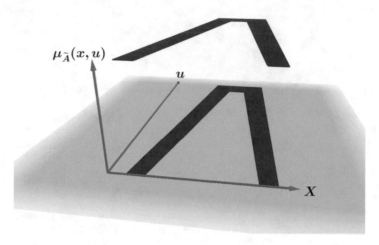

Fig. 2.21 Interval type-2 fuzzy set of Fig. 2.20b in space $\mathbb{R}^3$

2.2.1   Interval Type-2 FRBS

> *Type-2 fuzzy sets let us model the uncertainties that are inherent in words as well as other uncertainties. [31]*
>
> Jerry M. Mendel (2003)

The construction of the interval type-2 FRBS follows similar type-1 counterpart steps with the difference that, in the first step, which fuzzifies the independent inputs and outputs of the model to be analyzed, the linguistic variables that compose them are elaborated, with the help of specialists or information from the literature. At least one of the input linguistic variables is transformed into an interval type-2 fuzzy set. The block diagram of the interval type-2 FRBS is shown in Fig. 2.22. The interval type-2 FRBS consists of five components: input processor, inference, rule base, type-reducer, and output processor described in Fig. 2.22 and explained in the following.

1. **Input Processor**: Given $(x_1', x_2', \ldots, x_n') \in \mathbb{R}^n$ this block calculates the upper and lower membership degree of x_i', $i = 1, 2, \ldots, n$ for each linguistic variable.
2. **Rule Base**: The rule base of the interval type-2 FRBS has the same structure of the type-1 counterpart. The difference is the nature of the sets involved. Consider an interval type-2 FRBS with input $(x_1, x_2, \ldots, x_n)$ where x_i has the associated set defined by

$$T_{x_i} = \left\{ \widetilde{X}_{ij}, j \in I_i \right\}, \ i = 1, 2, \ldots, n, \tag{2.26}$$

where $I_i \subset \mathbb{N}$ is a set of indexes ordered from 1, $\widetilde{X}_{ij}$, are interval type-2 fuzzy sets defined in the universes $X_i, i = 1, 2, \ldots, n$ that represent the linguistic terms of the variable x_i. The output, $(y_1, y_2, \ldots, y_p)$ has an associated set

$$T_{y_s} = \left\{ \widetilde{Y}_s^k, k \in I_s \right\}, \ s = 1, 2, \ldots, p, \tag{2.27}$$

where $I_s \subset \mathbb{N}$ is a set of indexes ordered from 1, and $\widetilde{Y}_s^k$ is a type-2 set, defined in the universe $Y_s, s = 1, 2, \ldots, p$. Note that $\widetilde{Y}_s^k$ can be a point (singleton), an interval, or an interval type-2 fuzzy set. From now on and for sake of simplicity, when there is only one input it is denoted by x. Similarly, when there is only one output it is denoted by y. Furthermore, T_{y_1} is denote by T_y and $\widetilde{Y}_1^k$ by $\widetilde{Y}^k$.

The rule $R^m, m \in \mathbb{N}, \ m = 1, 2, \ldots, M$ of the interval type-2 FRBS is defined by

$$R^m : \text{ If } x_1 \text{ is } \widetilde{F}_1^m \text{ and } x_2 \text{ is } \widetilde{F}_2^m \text{ and } \ldots \text{ and } x_n \text{ is } \widetilde{F}_n^m \text{ then } y \text{ is } \widetilde{G}^m, \tag{2.28}$$

where $\widetilde{F}_k^m, \ k = 1, \ldots, n$ is an element of the set T_{x_i} is defined in Eq. (2.26) and $\widetilde{G}^m$ is an element of the set T_{y_s} is defined in Eq. (2.27).

3. **Fuzzy Inference**: the inference block performs the logical operations for interval type-2 FRBS.
4. **Type-Reducer**: the type-reducer block aims to use the Karnik–Mendel (KM) [23] algorithm that determines the minimum, y_L, and the maximum, y_R of all the centroids of type-1 fuzzy sets contained in the interval type-2 fuzzy set. There are other type-reducer methods that can be found for more details in [32].
5. **Defuzzification**: the defuzzified output of the interval type-2 FRBS is given by the average of y_L and y_R, that is,

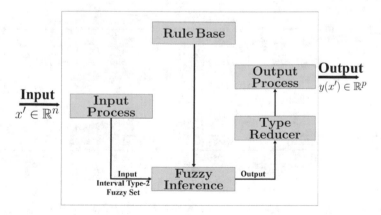

Fig. 2.22 Architecture of the interval type-2 fuzzy rule-based system [9, 10, 40]

$$y' = \frac{y_L + y_R}{2}, \tag{2.29}$$

noticing that y_L and y_R depend on each input crisp value.

2.2.1.1 Consequent of the Rules Base Is an Interval: Dongrui Wu's Method

Define the m rules R^m, $m = 1, 2, \ldots, M$ of the interval type-2 FRBS as in Eq. (2.28) where $\widetilde{F}_k^m$, $k = 1, \ldots, n$ is an element of the set T_{x_i} defined in (2.26) and $\widetilde{D}_0^m = [\underline{g}^m, \overline{g}^m]$. Giving the t-norm $\star$, which in this case is the **product** (see Example 2.1), and an input vector $(x_1', x_2', \ldots, x_n') \in \mathbb{R}^n$, the inference method starts calculating the upper and lower membership degrees of x_i' of the m-th rule, as follows:

$$\underline{\mu}_{\widetilde{F}_i^m}(x_i') \text{ and } \overline{\mu}_{\widetilde{F}_i^m}(x_i') \quad i = 1, 2, \ldots, n, \quad m = 1, 2, \ldots, M. \tag{2.30}$$

Next step is to determine $\underline{f}^m$ and $\overline{f}^m$ as

$$\underline{f}^m = \underline{\mu}_{\widetilde{F}_1^m}(x_1') \star \ldots \star \underline{\mu}_{\widetilde{F}_n^m}(x_n'), \quad m = 1, 2, \ldots, M, \tag{2.31}$$

and

$$\overline{f}^m = \overline{\mu}_{\widetilde{F}_1^m}(x_1') \star \ldots \star \overline{\mu}_{\widetilde{F}_n^m}(x_n'), \quad m = 1, 2, \ldots, M. \tag{2.32}$$

Thus, with the **KM** algorithm process [23], the following numbers can be calculated:

$$y_L = \min_{k \in [1, M-1]} \frac{\sum\limits_{m=1}^{k} \overline{f}^m \underline{g}^m + \sum\limits_{m=k+1}^{M} \underline{f}^m \underline{g}^m}{\sum\limits_{m=1}^{k} \overline{f}^m + \sum\limits_{m=k+1}^{M} \underline{f}^m} = \frac{\sum\limits_{m=1}^{L} \overline{f}^m \underline{g}^m + \sum\limits_{m=L+1}^{M} \underline{f}^m \underline{g}^m}{\sum\limits_{m=1}^{L} \overline{f}^m + \sum\limits_{m=L+1}^{M} \underline{f}^m}, \tag{2.33}$$

and

$$y_R = \max_{k \in [1, M-1]} \frac{\sum\limits_{m=1}^{k} \underline{f}^m \overline{g}^m + \sum\limits_{m=k+1}^{M} \overline{f}^m \overline{g}^m}{\sum\limits_{m=1}^{k} \underline{f}^m + \sum\limits_{m=k+1}^{M} \overline{f}^m} = \frac{\sum\limits_{m=1}^{R} \underline{f}^m \overline{g}^m + \sum\limits_{m=R+1}^{M} \overline{f}^m \overline{g}^m}{\sum\limits_{m=1}^{R} \underline{f}^m + \sum\limits_{m=R+1}^{M} \overline{f}^m}. \tag{2.34}$$

The points L and R are called switch points [32]. The point y_L is shown in Fig. 2.23a. Depicted in Fig. 2.23b is the point y_R.

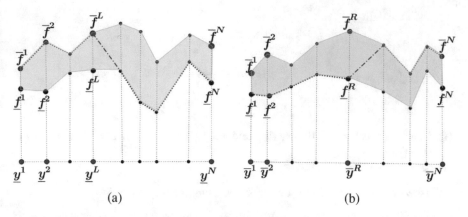

Fig. 2.23 Switch points L and R [9, 10, 49]

Finally, the interval type-2 FRBS defuzzification is the average of the values y_L and y_R as in Eq. (2.29). The details of the steps of the KM algorithm are explained in what follows.

Karnik–Mendel Algorithm

Consider the input vector $(x'_1, x'_2, \ldots, x'_n) \in \mathbb{R}^n$, calculate $\underline{f}^m$ and $\overline{f}^m$ as in Eq. (2.31) and Eq. (2.32) using Eq. (2.30). From there the following steps proceed.

(a) Sort in ascending order g^m, $m = 1, \ldots, M$, defined previously and associate their corresponding $[\underline{f}^m, \overline{f}^m]$, and rename it with the name of the previous order.

(b) Initialize f^m by setting,

$$f^m = \frac{\underline{f}^m + \overline{f}^m}{2}, \quad m = 1, 2, \ldots, M.$$

(c) Compute y' as being

$$y' = \frac{\sum\limits_{m=1}^{M} \underline{g}^m f^m}{\sum\limits_{m=1}^{M} f^m}.$$

(d) Find the switch point t k, $1 \leq k \leq M - 1$, such that $\underline{g}^k \leq y' \leq \underline{g}^{k+1}$.

(e) Define

$$f^m = \begin{cases} \overline{f}^m & \text{if } m \leq k, \\ \underline{f}^m & \text{if } m > k, \end{cases}$$

and calculate

$$y_l(k) = \frac{\sum\limits_{m=1}^{M} \underline{g}^m f^m}{\sum\limits_{m=1}^{M} f^m}.$$

(f) Check if $y_l(k) = y'$. If yes, stop and set $y_L = y_l(k)$ and $L = k$. If no, go to step (g).
(g) Set $y' = y_l(k)$ and go to step (c).

Likewise, the calculation is done for y_R, using the sequence $\overline{g}^m$.

Remark To further clarify step (a) of the y_L calculation, suppose that there are six rules and that: $\underline{g}^1 = 2$, $\underline{g}^2 = 1$, $\underline{g}^3 = 0$, $\underline{g}^4 = 5$, $\underline{g}^5 = 0$, and $\underline{g}^6 = 3$.
Then, ordering the sequence of $\underline{g}^m$ in ascending order and associating its corresponding $[\underline{f}^m, \overline{f}^m]$, it is obtained:

$$\underline{g}^3 = 0 \le \underline{g}^5 = 0 < \underline{g}^2 = 1 < \underline{g}^1 = 2 < y^6 = 3 < \underline{g}^4 = 5;$$
$$\parallel \qquad \parallel \qquad \parallel \qquad \parallel \qquad \parallel \qquad \parallel$$
$$\underline{g}^1 \qquad \underline{g}^2 \qquad \underline{g}^3 \qquad \underline{g}^4 \qquad \underline{g}^5 \qquad \underline{g}^6$$

$$[\underline{f}^3, \overline{f}^3] \; [\underline{f}^5, \overline{f}^5] \; [\underline{f}^2, \overline{f}^2] \; [\underline{f}^1, \overline{f}^1] \; [\underline{f}^6, \overline{f}^6] \; [\underline{f}^4, \overline{f}^4]$$
$$\parallel \qquad \parallel \qquad \parallel \qquad \parallel \qquad \parallel \qquad \parallel$$
$$[\underline{f}^1, \overline{f}^1] \; [\underline{f}^2, \overline{f}^2] \; [\underline{f}^3, \overline{f}^3] \; [\underline{f}^4, \overline{f}^4] \; [\underline{f}^5, \overline{f}^5] \; [\underline{f}^6, \overline{f}^6].$$

For a better understanding of the KM algorithm, the Example 2.11 is presented.

Example 2.11 It is considered the human reproduction rate of a population as the output of an interval type-2 FRBS to be built. This output consists of four intervals and the input variables are the fertility rate, x_1, and the country's economic growth rate, x_2. The input variable x_1 consists of two interval type-2 fuzzy sets, $\widetilde{X}_{11}$ and $\widetilde{X}_{12}$, which represent the linguistic terms low and high, respectively. Likewise, the variable x_2 is composed of two interval type-2 fuzzy sets $\widetilde{X}_{21}$ and $\widetilde{X}_{22}$, which also represent the linguistic terms low and high, respectively.

The fuzzy rule base is the following:

$$R^1 : \text{If } x_1 \text{ is } \widetilde{F}_1^1 = \widetilde{X}_{11} \text{ and } x_2 \text{ is } \widetilde{F}_2^1 = \widetilde{X}_{21} \text{ then } y \text{ is } G^1 = Y^1;$$

$$R^2 : \text{If } x_1 \text{ is } \widetilde{F}_1^2 = \widetilde{X}_{11} \text{ and } x_2 \text{ is } \widetilde{F}_2^2 = \widetilde{X}_{22} \text{ then } y \text{ is } G^2 = Y^2;$$

$$R^3 : \text{If } x_1 \text{ is } \widetilde{F}_1^3 = \widetilde{X}_{12} \text{ and } x_2 \text{ is } \widetilde{F}_2^3 = \widetilde{X}_{21} \text{ then } y \text{ is } G^3 = Y^3;$$

$$R^4 : \text{If } x_1 \text{ is } \widetilde{F}_1^4 = \widetilde{X}_{12} \text{ and } x_2 \text{ is } \widetilde{F}_2^4 = \widetilde{X}_{22} \text{ then } y \text{ is } G^4 = Y^4.$$

Table 2.3 Consequents of the interval type-2 FRBS with interval output

x_1	x_2 $\widetilde{X}_{21}$	$\widetilde{X}_{22}$
$\widetilde{X}_{11}$	$Y^1 = [\underline{y}^1, \overline{y}^1] = [0, 0.25]$	$Y^2 = [\underline{y}^2, \overline{y}^2] = [0, 0.25]$
$\widetilde{X}_{12}$	$Y^3 = [\underline{y}^3, \overline{y}^3] = [0.652, 1]$	$Y^4 = [\underline{y}^4, \overline{y}^4] = [0.1, 0.9]$

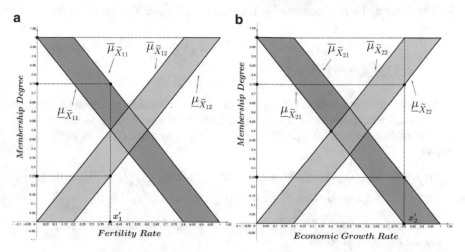

Fig. 2.24 FOU of the input variables: fertility rate and country's economic growth rate [9, 49]

The consequents Y^1, Y^2, Y^3, and Y^4 are defined in Table 2.3.

The shaded areas of the graph in Fig. 2.24a show the membership functions of input x_1: upper, $\overline{\mu}_{\widetilde{X}_{11}}$, and lower, $\underline{\mu}_{\widetilde{X}_{11}}$, for the fuzzy set $\widetilde{X}_{11}$, as well as the membership functions: upper, $\overline{\mu}_{\widetilde{X}_{12}}$, and lower, $\underline{\mu}_{\widetilde{X}_{12}}$, for the fuzzy set $\widetilde{X}_{12}$. Depicted in Fig. 2.24b are the membership functions of input x_2: upper $\overline{\mu}_{\widetilde{X}_{21}}$ and lower $\underline{\mu}_{\widetilde{X}_{21}}$, corresponding to the fuzzy set $\widetilde{X}_{21}$; and the membership functions: upper $\overline{\mu}_{\widetilde{X}_{22}}$, and lower, $\underline{\mu}_{\widetilde{X}_{22}}$, for the fuzzy set $\widetilde{X}_{22}$ [20, 49]. The universes for the variables x_1 and x_2 are $X_1 = X_2 = [0, 1]$.

The upper and lower membership functions of $\widetilde{X}_{11}$ and $\widetilde{X}_{12}$ are defined as

$$\underline{\mu}_{\widetilde{X}_{11}}(x) = \begin{cases} 1 - \dfrac{x}{0.8} & \text{if } 0 \le x < 0.8, \\[2mm] 0 & \text{otherwise,} \end{cases} \qquad \overline{\mu}_{\widetilde{X}_{11}}(x) = \begin{cases} 1 & \text{if } 0 \le x < 0.2, \\[2mm] \dfrac{1-x}{0.8} & \text{if } 0.2 \le x \le 1, \\[2mm] 0 & \text{otherwise,} \end{cases}$$

$$\underline{\mu}_{\widetilde{X}_{12}}(x) = \begin{cases} \dfrac{x - 0.2}{0.8} & \text{if } 0.2 \le x \le 1, \\ \\ 0 & \text{otherwise,} \end{cases} \qquad \overline{\mu}_{\widetilde{X}_{12}}(x) = \begin{cases} \dfrac{x}{0.8} & \text{if } 0 \le x < 0.8, \\ 1 & \text{if } 0.8 \le x \le 1. \\ 0 & \text{otherwise.} \end{cases}$$

The upper and lower membership functions of $\widetilde{X}_{21}$ and $\widetilde{X}_{22}$ are given by

$$\underline{\mu}_{\widetilde{X}_{21}}(x) = \underline{\mu}_{\widetilde{X}_{11}}(x), \ \overline{\mu}_{\widetilde{X}_{21}}(x) = \overline{\mu}_{\widetilde{X}_{11}}(x), \ \underline{\mu}_{\widetilde{X}_{22}}(x) = \underline{\mu}_{\widetilde{X}_{12}}(x),$$

and $\overline{\mu}_{\widetilde{X}_{22}}(x) = \overline{\mu}_{\widetilde{X}_{12}}(x).$

Consider an input vector, $(x_1', x_2') = (0.4, \ 0.8)$. Hereafter is the calculation of the intervals whose extremes are the upper and lower membership functions at the point x' of the four interval type-2 fuzzy sets, components of the two input variables, namely:

$$[\underline{\mu}_{\widetilde{X}_{11}}(x_1'), \overline{\mu}_{\widetilde{X}_{11}}(x_1')] = [0.5, \ 0.75]; \quad [\underline{\mu}_{\widetilde{X}_{12}}(x_1'), \overline{\mu}_{\widetilde{X}_{12}}(x_1')] = [0.25, \ 0.5];$$
$$[\underline{\mu}_{\widetilde{X}_{21}}(x_2'), \overline{\mu}_{\widetilde{X}_{21}}(x_2')] = [0, \ 0.25]; \quad [\underline{\mu}_{\widetilde{X}_{22}}(x_2'), \overline{\mu}_{\widetilde{X}_{22}}(x_2')] = [0.75, \ 1].$$

Next, it is calculated $[\underline{f}^m, \overline{f}^m]$ for $m = 1, \ 2, \ 3, \ 4$, corresponding to each rule using the $t-$norm, **product**:

$$[\underline{f}^1, \overline{f}^1] = [\underline{\mu}_{\widetilde{X}_{11}}(x_1') \cdot \underline{\mu}_{\widetilde{X}_{21}}(x_2'), \overline{\mu}_{\widetilde{X}_{11}}(x_1') \cdot \overline{\mu}_{\widetilde{X}_{21}}(x_2')] = \ [0, \ 0.1875];$$
$$[\underline{f}^2, \overline{f}^2] = [\underline{\mu}_{\widetilde{X}_{11}}(x_1') \cdot \underline{\mu}_{\widetilde{X}_{22}}(x_2'), \overline{\mu}_{\widetilde{X}_{11}}(x_1') \cdot \overline{\mu}_{\widetilde{X}_{22}}(x_2')] = [0.375, \ 0.75];$$
$$[\underline{f}^3, \overline{f}^3] = [\underline{\mu}_{\widetilde{X}_{12}}(x_1') \cdot \underline{\mu}_{\widetilde{X}_{21}}(x_2'), \overline{\mu}_{\widetilde{X}_{12}}(x_1') \cdot \overline{\mu}_{\widetilde{X}_{21}}(x_2')] = \ [0, \ 0.125];$$
$$[\underline{f}^4, \overline{f}^4] = [\underline{\mu}_{\widetilde{X}_{12}}(x_1') \cdot \underline{\mu}_{\widetilde{X}_{22}}(x_2'), \overline{\mu}_{\widetilde{X}_{12}}(x_1') \cdot \overline{\mu}_{\widetilde{X}_{22}}(x_2')] = [0.1875, \ 0.5].$$

Shown in Table 2.4 is the activation interval and its respective consequent for each rule.

Applying the KM algorithm is found y_L compute as shown in the next steps:

(a) Sort $\underline{y}^m, m = 1, 2, 3, 4$ in increasing order and rename them so that:

$$\underline{y}^1 = 0 \le \underline{y}^2 = 0 \le \underline{y}^3 = 0.1 \le \underline{y}^4 = 0.652.$$

Table 2.4 Activation interval of each rule and their corresponding consequent for Wu inference method

Rule $n^{\underline{o}}$	Activation interval	$\Rightarrow$	Consequent
R^1	$[\underline{f}^1, \overline{f}^1] = [0, \ 0.1875]$	$\Rightarrow$	$[\underline{y}^1, \overline{y}^1] = [0, 0.25]$
R^2	$[\underline{f}^2, \overline{f}^2] = [0.375, \ 0.75]$	$\Rightarrow$	$[\underline{y}^2, \overline{y}^2] = [0, \ 0.25]$
R^3	$[\underline{f}^3, \overline{f}^3] = [0, \ 0.125]$	$\Rightarrow$	$[\underline{y}^3, \overline{y}^3] = [0.652, \ 1]$
R^4	$[\underline{f}^4, \overline{f}^4] = [0.1875, \ 0.5]$	$\Rightarrow$	$[\underline{y}^4, \overline{y}^4] = [0.1, \ 0.9]$

(b) Determine f^m defined by $f^m = \dfrac{\underline{f}^m + \overline{f}^m}{2}$, for the corresponding ordered values for $m = 1, 2, 3, 4$.

$$f^1 = \frac{0 + 0.1875}{2} = 0.09375; \qquad f^2 = \frac{0.375 + 0.75}{2} = 0.5625;$$

$$f^3 = \frac{0.1875 + 0.56}{2} = 0.34375; \qquad f^4 = \frac{0 + 0.125}{2} = 0.0625.$$

Next calculate, y' as

$$y' = \frac{\displaystyle\sum_{m=1}^{4} y^m f^m}{\displaystyle\sum_{n=1}^{4} f^m} = \frac{y^1 f^1 + y^2 f^2 + y^3 f^3 + y^4 f^4}{f^1 + f^2 + f^3 + f^4}$$

$$= \frac{(0)(0.09375) + (0)(0.5625) + (0.1)(0.34375) + (0.652)(0.0625)}{0.09375 + 0.5625 + 0.34375 + 0.0625}$$

$$= 0.07071.$$

(c) Find the switch point such that $\underline{y}^k \le y' \le \underline{y}^{k+1}$, being k the highest index that verifies those inequalities. Note that, $\underline{y}^1 = \underline{y}^2 \le y' \le \underline{y}^3$, therefore $k = 2$.

(d) Define $f^m = \begin{cases} \overline{f}^m & \text{if } m \le 2, \\ \underline{f}^m & \text{if } m > 2. \end{cases}$ for the reordered indexes n. Therefore,

$$\overline{f}^1 = 0.1875, \quad \overline{f}^2 = 0.75, \quad \underline{f}^3 = 0.1875, \quad \underline{f}^4 = 0.$$

Following, calculate $y_l(k)$:

$$y_l(k) = \frac{y^1 \overline{f}^1 + y^2 \overline{f}^2 + y^3 \underline{f}^3 + y^4 \underline{f}^4}{\overline{f}^1 + \overline{f}^2 + \underline{f}^3 + \underline{f}^4}$$

$$= \frac{(0)(0.1875) + (0)(0.75) + (0.1)(0.1875) + (0.652)(0)}{0.1875 + 0.75 + 0.1875 + 0}$$

$$= \frac{0.01875}{1.125} = 0.01666666667.$$

(e) As $y_l(k) \ne y'$ then continue the process.

(f) Define $y' = y_l(k)$ and find the switch point. Note that, $\underline{y}^2 \le y' \le \underline{y}^3$, that is, $(0 \le 0.01666666667 \le 0.1)$. Thus, $k = 2$.

Define, $f^m = \begin{cases} \overline{f}^m & \text{if } m \le 2, \\ \underline{f}^m & \text{if } m > 2, \end{cases}$ and calculate $y_l(k)$:

$$y_l(k) = \frac{y^1 \overline{f}^1 + y^2 \overline{f}^2 + y^3 \underline{f}^3 + y^4 \underline{f}^4}{\overline{f}^1 + \overline{f}^2 + \underline{f}^3 + \underline{f}^4}$$

$$= \frac{(0)(0.1875) + (0)(0.75) + (0.1)(0.1875) + (0.652)(0)}{0.1875 + 0.75 + 0.1875 + 0}$$

$$= \frac{0.01875}{1.125} \approx 0.01667.$$

Since $y' = y_l(k)$ then $y_L \approx 0.01667$ and $L = 2$.

Next, it is calculated y_R using an analogous procedure. These steps are repeated in what follows with the aim to reinforce the methodology.

(a) Sort in ascending order $\overline{y}^m$, $m = 1, 2, 3, 4$. Thus, $\overline{y}^1 \le \overline{y}^2 \le \overline{y}^3 \le \overline{y}^4$, that is, $0.25 \le 0.25 \le 0.9 \le 1$.

(b) Find $f^1 = 0.09375$, $f^2 = 0.5625$, $f^3 = 0.34375$, $f^4 = 0.0625$.
    As aforesaid, calculate y':

$$y' \frac{(0.25)(0.09375) + (0.25)(0.5625) + (0.9)(0.34375) + (1)(0.0625)}{0.09375 + 0.5625 + 0.34375 + 0.0625}$$

$$= 0.50441.$$

(c) Find the switch point noticing that, $\overline{y}^2 \le y' \le \overline{y}^3$ to conclude that $k = 2$.

(d) Define $f^m = \begin{cases} \underline{f}^m & \text{if } m \le 2, \\ \overline{f}^m & \text{if } m > 2. \end{cases}$ Calculate $y_r(k)$:

$$y_r(k) = \frac{\overline{y}^1 \underline{f}^1 + \overline{y}^2 \underline{f}^2 + \overline{y}^3 \overline{f}^3 + \overline{y}^4 \overline{f}^4}{\underline{f}^1 + \underline{f}^2 + \overline{f}^3 + \overline{f}^4}$$

$$= \frac{(0.25)(0) + (0.25)(0.375) + (1)(0.25) + (0.9)(0.5)}{0 + 0.375 + 0.125 + 1.5} = 0.66875.$$

(e) As $y_r(k) \ne y'$ then continue the process.

(f) Define $y' = y_r(k)$ and find the switch point. Note that, $\underline{y}^2 \le y' \le \underline{y}^3$, that is, $(0.25 \le 0.66875 \le 0.9)$ and, as a consequence, $k = 2$.

Define, $f^m = \begin{cases} \overline{f}^m & \text{if } m \le 2, \\ \underline{f}^m & \text{if } m > 2, \end{cases}$ and calculate $y_r(k)$:

$$y_r(k) = \frac{\overline{y}^1 \underline{f}^1 + \overline{y}^2 \underline{f}^2 + \overline{y}^3 \overline{f}^3 + \overline{y}^4 \overline{f}^4}{\underline{f}^1 + \underline{f}^2 + \overline{f}^3 + \overline{f}^4}$$

$$= \frac{(0.25)(0) + (0.25)(0.375) + (1)(0.25) + (0.9)}{=} 0.66875.$$

Being $y' = y_r(k)$ then $y_R = 0.66875$ and $R = 2$.

Finally, the defuzzified output is given by $y' = \dfrac{y_L + y_R}{2} = 0.342725$.

As a conclusion, for the values 0.4 of fertility rate and 0.8 of the country's economic growth rate, it is obtained a corresponding approximately 0.34 of reproduction rate. The previous calculations have been based on [49] and detailed in [9, 10].

Exercise 2.6 Based on Example 2.11, determine the output, from the crisp input $(x_1', x_2') = (0.2, 0.6)$.

2.2.1.2 Consequent Is a Function of the Input Variables: Takagi–Sugeno–Kang Inference Method

As defined in Eq. (2.1), a linear combination of type-1 interval fuzzy numbers is a type-1 fuzzy set. Consider $(x_1, x_2, \ldots, x_n)$ the i-th rule of m rules, $m = 1, 2, \ldots, M$, can be expressed as

$$R^m : \text{If } x_1 \text{ is } \widetilde{F}_1^m \text{ and } x_2 \text{ is } \widetilde{F}_2^m \text{ and } \ldots \text{ and } x_n \text{ is } \widetilde{F}_n^m \text{ then } y \text{ is } \widetilde{G}^m$$

$$= \lambda^m(x_1, x_2, \ldots, x_n), \tag{2.35}$$

where λ^m is a function of the variables $(x_1, x_2, \ldots, x_n)$. In particular, in the applications of this book, the function λ^m is used as

$$\lambda^m(x_1, x_1, \ldots, x_n) = D_0^m, \tag{2.36}$$

or linear of the form

$$\lambda^m(x_1, x_1, \ldots, x_n) = D_0^m + D_1^m x_1 + \ldots + D_n^m x_n, \tag{2.37}$$

where $m = 1, \ldots, M$, and D_j^m, $j = 0, \ldots, n$ are intervals (interpreted as type-1 interval fuzzy numbers).

Hereafter it is examined Eq. (2.37) in the case that the output is a linear combination of intervals, interpreted as type-1 interval fuzzy numbers [32].

Consider an interval type-2 FRBS with rules R^m, $m = 1, 2, \ldots, M$ defined as in Eq. (2.35), where Y^m is the output of the m−th rule, it is also a type-1 fuzzy set (because it is a linear combination of type-1 fuzzy sets); and $\widetilde{F}_i^m$, $i = 1, \ldots, n$, are fuzzy antecedent sets of type-2. In this case, the output is expressed by

$$Y^m = D_0^m + D_1^m x_1 + \ldots + D_n^m x_n, \qquad D_j^m = [c_j^m - a_j^m, c_j^m + a_j^m],$$

wherein c_j^m, are centers and a_j^m amplitudes of the intervals D_j^m, $m = 1, 2, \ldots, M$, $j = 0, 1, \ldots, n,$.

Consider a $t-$norm $\star$, **minimum** or **product** (see Example 2.1), an input vector $(x_1', x_2', \ldots, x_n') \in \mathbb{R}^n$ and the upper and lower membership degrees of x_i' corresponding to the m-th rule, $m = 1, 2, \ldots, M$ (see Eq. (2.30)). Then, are calculated

$$\underline{f}^m = \underline{\mu}_{\tilde{F}_1^m}(x_1') \star \ldots \star \underline{\mu}_{\tilde{F}_n^m}(x_n'), \quad \overline{f}^m = \overline{\mu}_{\tilde{F}_1^m}(x_1') \star \ldots \star \overline{\mu}_{\tilde{F}_1^m}(x_n'),$$

The consequent is given by $Y^m = [\underline{y}^m, \overline{y}^m]$, that is expressed by

$$\underline{y}^m = \sum_{i=1}^{n} c_i^m x_i' + c_0^m - \left(\sum_{i=1}^{n} a_i^m |x_i'| + a_0^m \right); \qquad (2.38)$$

$$\overline{y}^m = \sum_{i=1}^{n} c_i^m x_i' + c_0^m + \left(\sum_{i=1}^{n} a_i^m |x_i'| + a_0^m \right). \qquad (2.39)$$

Using the KM algorithm [33], it is determined, as in the equations Eq. (2.33) and Eq. (2.34), the values y_L and y_R as

$$y_L = \min_{k \in [1, M-1]} \frac{\sum_{m=1}^{k} \overline{f}^m \underline{y}^m + \sum_{m=k+1}^{M} \underline{f}^m \underline{y}^m}{\sum_{m=1}^{k} \overline{f}^m + \sum_{m=k+1}^{M} \underline{f}^m} = \frac{\sum_{m=1}^{L} \overline{f}^m \underline{y}^m + \sum_{m=L+1}^{M} \underline{f}^m \underline{y}^m}{\sum_{m=1}^{L} \overline{f}^m + \sum_{m=L+1}^{M} \underline{f}^m}, \qquad (2.40)$$

$$y_R = \max_{k \in [1, M-1]} \frac{\sum_{m=1}^{k} \underline{f}^m \overline{y}^m + \sum_{m=k+1}^{M} \overline{f}^m \overline{y}^m}{\sum_{m=1}^{k} \underline{f}^m + \sum_{m=k+1}^{M} \overline{f}^m} = \frac{\sum_{m=1}^{R} \underline{f}^m \overline{y}^m + \sum_{m=R+1}^{M} \overline{f}^m \overline{y}^m}{\sum_{m=1}^{R} \underline{f}^m + \sum_{m=R+1}^{M} \overline{f}^m}. \qquad (2.41)$$

Defuzzifier The defuzzified value is given by $y' = \dfrac{y_L + y_R}{2}$, calculated at the corresponding value x'.

The aim of the Example 2.12 is to illustrate the method presented in this section.

Example 2.12 Inspired by the advertising poster [39] shown in Fig. 2.25, a fictitious model of an interval type-2 FRBS using an interval type-2 TSK inference method has been built, taking information from an ANFIS training for a type-1 FRBS.

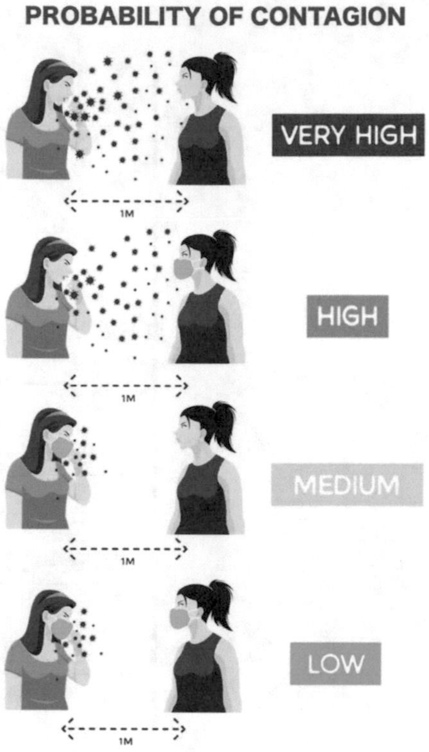

Fig. 2.25 This advertisement is from the Mauritian company, Remark Air©. The word "probability" should be mathematically understood as "possibility." Image kindly provided from the source: https://remask.online/can-a-simple-face-mask-help-combat-covid-19-yes-only-if-everyone-wears-a-mask/

Suppose there are 100 people gathered in the same environment, among these, 50 people are transmitters of the COVID-19. People of the group may be wearing masks or not. The input variable of the interval type-2 FRBS is the number of people wearing masks (x), whose domain is [0, 100], with the linguistic terms Small $(\widetilde{S})$, Medium $(\widetilde{M})$, Large $(\widetilde{L})$, and Very Large $(\widetilde{VL})$. The output variable is the contamination risk (y), varying in the range [0.1, 0.9]. The choice of this interval comes from the fact that, even if everyone uses a mask, there is still a possibility of infection and, even if none of the people uses a mask, contamination may not occur. The output consists of functions that depend on the input variables. It is determined the risk of contamination for 19 people wearing a mask.

The fuzzy rules are presented in the following:

R^1 : If x is $\widetilde{S}$ then y is $Y^1 = [0.8994, 0.9794] + [-0.01124, -0.00524]x$;

R^2 : If x is $\widetilde{M}$ then y is $Y^2 = [0.9284, 0.9304] + [-0.009723, -0.005723]x$;

R^3 : If x is $\widetilde{L}$ then y is $Y^3 = [0.9022, 0.9042] + [-0.008764, -0.006764]x$;

R^4 : If x is $\widetilde{VL}$ then y is $Y^4 = [0.9019, 0.9039] + [-0.00806, -0.00646]x$.

The corresponding upper and lower membership functions are:

$$\overline{\mu}_{\widetilde{S}}(x) = \begin{cases} 1 & \text{if } 0 \leq x < 10, \\ \dfrac{-x + 23.5}{13.5} & \text{if } 10 \leq x < 23.5, \\ 0 & \text{otherwise,} \end{cases}$$

$$\underline{\mu}_{\widetilde{S}}(x) = \begin{cases} 0.95 & \text{if } 0 \leq x < 7.11, \\ \dfrac{-0.95x + 17.955}{11.79} & \text{if } 7.11 \leq x < 18.9, \\ 0 & \text{otherwise,} \end{cases}$$

$$\overline{\mu}_{\widetilde{M}}(x) = \begin{cases} \dfrac{x - 9.983}{13.417} & \text{if } 9.983 \leq x < 23.4, \\ 1 & \text{if } 23.4 \leq x < 43.4, \\ \dfrac{-x + 56.8}{13.4} & \text{if } 43.4 \leq x < 56.8, \\ 0 & \text{otherwise,} \end{cases}$$

$$\underline{\mu}_{\widetilde{M}}(x) = \begin{cases} \dfrac{0.95x - 14.345}{11.3} & \text{if } 15.1 \leq x < 26.4, \\ 0.95 & \text{if } 26.4 \leq x < 39.5, \\ \dfrac{-0.95x + 48.735}{11.8} & \text{if } 39.5 \leq x < 51.3, \\ 0 & \text{otherwise,} \end{cases}$$

$$\overline{\mu}_{\widetilde{L}}(x) = \begin{cases} \dfrac{x - 43.32}{12.98} & \text{if } 43.32 \leq x < 56.3, \\ 1 & \text{if } 56.3 \leq x < 77, \\ \dfrac{-x + 90.3}{13.3} & \text{if } 77 \leq x < 90.3, \\ 0 & \text{otherwise,} \end{cases}$$

$$\underline{\mu}_{\widetilde{L}}(x) = \begin{cases} \dfrac{0.95x - 47.025}{9} & \text{if } 49.5 \leq x < 58.5, \\ 0.95 & \text{if } 58.5 \leq x < 72.4, \\ \dfrac{-0.95x + 81.035}{12.9} & \text{if } 72.4 \leq x < 85.3, \\ 0 & \text{otherwise,} \end{cases}$$

$$\overline{\mu}_{\widetilde{VL}}(x) = \begin{cases} \dfrac{x - 76.65}{13.35} & \text{if } 76.65 \leq x < 90, \\ 1 & \text{if } 90 \leq x \leq 100, \\ 0 & \text{otherwise,} \end{cases}$$

$$\underline{\mu}_{\widetilde{VL}}(x) = \begin{cases} \dfrac{0.95x - 76.8835}{12.6} & \text{if } 80.93 \leq x < 93.53, \\ 0.95 & \text{if } 93.53 \leq x \leq 100, \\ 0 & \text{otherwise.} \end{cases}$$

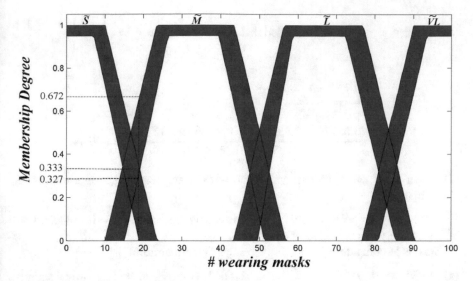

Fig. 2.26 FOU of the input variable

Consider the input value $x' = 19$ as shown in Fig. 2.26. Then, calculating $\underline{f}^m$ and $\overline{f}^m$, for $m = 1, 2$, as in Eqs. (2.31) and (2.32), it is obtained

$$[\underline{f}^1, \overline{f}^1] = [\underline{\mu}_{\tilde{S}}(x'_1), \overline{\mu}_{\tilde{S}}(x'_1)] = [0,\ 0.333]$$
$$[\underline{f}^2, \overline{f}^2] = [\underline{\mu}_{\tilde{M}}(x'_1), \overline{\mu}_{\tilde{M}}(x'_1)] = [0.32787,\ 0.67206].$$

Furthermore, it is calculated $\underline{y}^m$ and $\overline{y}^m$, for $m = 1, 2$, as in Eq. (2.38) and Eq. (2.39), to obtain

$$\underline{y}^1 = 0.685764, \quad \overline{y}^1 = 0.879764, \quad \underline{y}^2 = 0.743663, \quad and \quad \overline{y}^2 = 0.821663.$$

Using the KM algorithm [23], it is determined y_L as described hereafter.

(a) Being $\underline{y}^1 = 0.685764 \le \underline{y}^2 = 0.743663$, the steps of the algorithm follow.

(b) Calculate $f^1 = \dfrac{\underline{f}^1 + \overline{f}^1}{2} = 0.49995$, $f^2 = \dfrac{\underline{f}^2 + \overline{f}^2}{2} = 0.166665$, thus it is obtained

$$y' = \frac{(0.685764)(0.49995) + (0.743663)(0.166665)}{0.49995 + 0.166665} = 0.7002397.$$

(c) As $\underline{y}^1 \leq y' \leq \underline{y}^2$, it is concluded that $k = 1$. Therefore $f^m = \begin{cases} \overline{f}^m & \text{if } m \leq 1, \\ \underline{f}^m & \text{if } m > 1, \end{cases}$

and

$$y_l(1) = \frac{\underline{y}^1 \overline{f}^1 + \underline{y}^2 f^2}{\overline{f}^1 + f^2}$$
$$= \frac{(0.685764)(0.3333) + (0.743663)(0.32787)}{0.3333 + 0.32787} = 0.71447574.$$

(d) Being $y_l(k) \neq y'$, rename $y' = 0.71447574$ and restart the algorithm as in item (c).

In fact, as $\underline{y}^1 \leq y' \leq \underline{y}^2$ and $k = 1$, it is concluded that $y_l(k) = 0.71447574 = y'$. Therefore $y_L = 0.71447574$ and $L = 1$.

Next, is to compute y_R following the steps of the algorithm.

(a) As $\overline{y}^1 = 0.879764 > \overline{y}^2 = 0.8216663$, sort y^m, $m = 1.2$, in ascending order, and having $\overline{y}^1 = 0.8216663$ and $\overline{y}^2 = 0.879764$, and their corresponding $[\underline{f}^1, \overline{f}^1] = [0.32787, 0.67206]$, $[\underline{f}^2, \overline{f}^2] = [0, 0.3333]$, respectively.

(b) Calculate $f^1 = \frac{\underline{f}^1 + \overline{f}^1}{2} = 0.49995$, $f^2 = \frac{\underline{f}^2 + \overline{f}^2}{2} = 0.166665$, obtaining

$$y' = \frac{(0.821663)(0.49995) + (0.879764)(0.166665)}{0.49995 + 0.166665} = 0.8361142.$$

(c) As $\overline{y}^1 \leq y' \leq \overline{y}^2$, it is concluded that $k = 1$. Therefore $f^m = \begin{cases} \underline{f}^m & \text{if } m \leq 1, \\ \overline{f}^m & \text{if } m > 1, \end{cases}$

and

$$y_r(k) = \frac{\overline{y}^1 \underline{f}^1 + \overline{y}^2 \overline{f}^2}{\underline{f}^1 + \overline{f}^2}$$
$$= \frac{(0.0.821663)(0.32787) + (0.879764)(0.3333)}{0.32787 + 0.3333} = 0.85080085.$$

(d) Being $y_r(k) \neq y'$, rename $y' = 0.85080085$ and restart the algorithm as in item c).

Concluding that $k = 1$ and $y_r(k) = 0.85080085 = y'$. Therefore $y_R = 0.85080085$ and $R = 1$.

Finally, the defuzzification value is given by $y' = \frac{y_L + y_R}{2} = 0.782638$.

The conclusion of the calculations is that for $x' = 19$, namely, the number of people wearing a mask, then the risk of contamination is approximately 0.78.

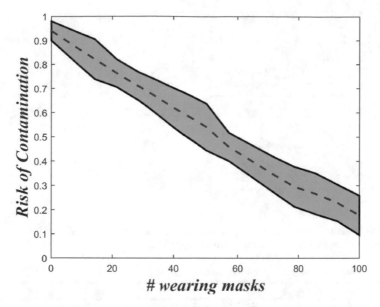

Fig. 2.27 The output of the interval type-2 FRBS for $x' \in [0, \ 100]$

Calculations made for all values of $x' \in [0, \ 100]$ determine the output of the interval type-2 FRBS, shown in Fig. 2.27. Detailing this figure, the graphs of the functions defined in $[0, \ 100]$ are: the output y_R (upper function), the output y_L (lower function), and the defuzzification y' (central function).

Exercise 2.7 Using the model of Example 2.12, determine the risk of contamination for 72 people who are wearing a mask.

2.2.1.3 Consequent of the Rule Base Is an Interval Type-2 Fuzzy Set: Mamdani Inference Method

Consider an interval type-2 FRBS with input $(x_1, x_2, \ldots, x_n)$ where x_i has the associated set $T_{x_i}, i = 1, 2, \ldots, n$, as in Eq. (2.26), $I_i \subset \mathbb{N}$ is a set of indexes ordered from 1, with Mamdani inference method in which the output, $(y_1, y_2, \ldots, y_p)$, is associated with $T_{y_s}, s = 1, 2, \ldots, p$ as defined in Eq. (2.27), where $\tilde{Y}_s^k$ is an interval type-2 fuzzy set, for all $k \in I_s$. The m rules R^m, $m = 1, 2, \ldots, M$ of the interval type-2 FRBS are defined as in Eq. (2.28).

Depicted in Fig. 2.28 is a block diagram of this method with the input (x_1, x_2) and the output y. For simplicity, consider two rules R^1 and R^2, with two antecedents and one consequent given by an interval type-2 fuzzy set, each one of the rules is given by

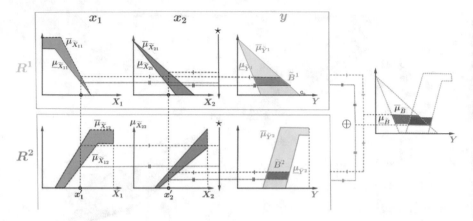

Fig. 2.28 Mamdani inference method for an interval type-2 fuzzy set utilizing the $\star$ t-norm of **minimum** and the $\oplus$, the s–norm **maximum** [40] and [8]

Table 2.5 Activation degree of the fuzzy rules

Rule $n^{\underline{o}}$	Activation degree		$\Rightarrow$	Consequent
R^1	$\underline{f}^1 = \underline{\mu}_{\widetilde{X}_{11}}(x_1') \star \underline{\mu}_{\widetilde{X}_{21}}(x_2')$	$\overline{f}^1 = \overline{\mu}_{\widetilde{X}_{11}}(x_1') \star \overline{\mu}_{\widetilde{X}_{21}}(x_2')$	$\Rightarrow$	$\widetilde{Y}^1$
R^2	$\underline{f}^2 = \underline{\mu}_{\widetilde{X}_{12}}(x_1') \star \underline{\mu}_{\widetilde{X}_{22}}(x_2')$	$\overline{f}^2 = \overline{\mu}_{\widetilde{X}_{12}}(x_1') \star \overline{\mu}_{\widetilde{X}_{22}}(x_2')$	$\Rightarrow$	$\widetilde{Y}^2$

$$R^1 : \text{If } x_1 \text{ is } \widetilde{X}_{11} \text{ and } x_2 \text{ is } \widetilde{X}_{21} \text{ then } y \text{ is } \widetilde{Y}^1;$$

$$R^2 : \text{If } x_1 \text{ is } \widetilde{X}_{12} \text{ and } x_2 \text{ is } \widetilde{X}_{22} \text{ then } y \text{ is } \widetilde{Y}^2.$$

The calculation utilizes the input vector (x_1', x_2').

First, the upper and lower activation degree of the rules, R^1 and R^2, are calculated, as shown in Table 2.5.

Afterwards, if $y \in Y$ then the interval type-2 fuzzy sets $\widetilde{B}^1$ and $\widetilde{B}^2$ are obtained, with the upper and lower membership functions given by:

$$\overline{\mu}_{\widetilde{B}^1}(y) = \overline{f}^1 \star \overline{\mu}_{\widetilde{Y}^1}(y) \qquad \underline{\mu}_{\widetilde{B}^1}(y) = \underline{f}^1 \star \underline{\mu}_{\widetilde{Y}^1}(y)$$

$$\overline{\mu}_{\widetilde{B}^2}(y) = \overline{f}^2 \star \overline{\mu}_{\widetilde{Y}^2}(y) \qquad \underline{\mu}_{\widetilde{B}^2}(y) = \underline{f}^2 \star \underline{\mu}_{\widetilde{Y}^2}(y).$$

Finally, the interval type-2 fuzzy set, $\widetilde{B}$, resulting from the inference is given by the following upper and lower membership functions, respectively:

$$\overline{\mu}_{\widetilde{B}}(y) = \overline{\mu}_{\widetilde{B}^1}(y) \oplus \overline{\mu}_{\widetilde{B}^2}(y) \quad \text{and} \quad \underline{\mu}_{\widetilde{B}}(y) = \underline{\mu}_{\widetilde{B}^1}(y) \oplus \underline{\mu}_{\widetilde{B}^2}(y).$$

The interval type-2 FRBS process continues with the defuzzification from the Mamdani inference method. In this step, the generalized centroid concept is explained.

Fig. 2.29 Embedded type-1
fuzzy set $A_e(l)$

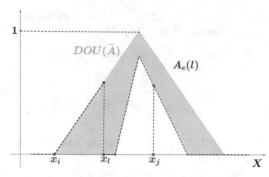

Fig. 2.30 Embedded type-1
fuzzy set $A_e(r)$

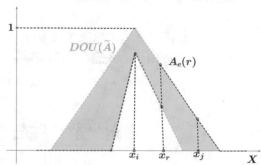

Generalized Centroid of an Interval Type-2 Fuzzy Set

Definition 2.16 The type-1 fuzzy set, A_e, defined by the membership μ_{A_e} is said
to be embedded in $\tilde{A}$ if,

$$\underline{\mu}_{\tilde{A}}(x) \leq \mu_{A_e}(x) \leq \overline{\mu}_{\tilde{A}}(x), \quad \text{for all } x \in X.$$

Denote by $E_{\tilde{A}}$ the set of all embedded sets A_e in $\tilde{A}$.

Shown in Figs. 2.29 and 2.30 are type-1 fuzzy sets, $A_e(l)$ and $A_e(r)$, respectively,
embedded in the interval type-2 fuzzy set $\tilde{A}$.

Discrete Version of the Generalized Centroid

The version used in Chap. 3 applications of the generalized centroid is discrete, that
is, $X = \{x_1 < x_2 < \ldots < x_n\}$ is a universe of discourse. Its calculation is justified
in the following:

Definition 2.17 Given an interval type-2 fuzzy set, $\tilde{A}$, defined in X, consider two
specific embedded subsets [34] of $\tilde{A}$, $A_e(l)$, defined as

$$\mu_{A_e(l)}(x_i) = \begin{cases} \overline{\mu}_{\tilde{A}}(x_i) \text{ if } i \leq l \\ \underline{\mu}_{\tilde{A}}(x_i) \text{ if } i > l, \end{cases}$$

where $i = 1, 2, \ldots, n$ and l is the so-called switch point of $A_e(l)$. A graphic of this type is shown in Fig. 2.29.

Likewise, consider fuzzy set $A_e(r)$ defined by

$$\mu_{A_e(r)}(x_i) = \begin{cases} \underline{\mu}_{\widetilde{A}}(x_i) \text{ if } i \leq r \\ \overline{\mu}_{\widetilde{A}}(x_i) \text{ if } i > r, \end{cases}$$

where $i = 1, 2, \ldots, n$ and r is the so-called switch point of $A_e(r)$. A graph of this type is shown in Fig. 2.30.

Then, it has been demonstrated [33] that exist L and R such that

$$y_L(\widetilde{A}) = \min_{l\in[1,n-1]} \frac{\sum\limits_{i=1}^{l} x_i\overline{\mu}_{\widetilde{A}}(x_i) + \sum\limits_{i=l+1}^{n} x_i\underline{\mu}_{\widetilde{A}}(x_i)}{\sum\limits_{i=1}^{l} \overline{\mu}_{\widetilde{A}}(x_i) + \sum\limits_{i=l+1}^{n} \underline{\mu}_{\widetilde{A}}(x_i)}$$

$$= \frac{\sum\limits_{i=1}^{L} x_i\overline{\mu}_{\widetilde{A}}(x_i) + \sum\limits_{i=L+1}^{n} x_i\underline{\mu}_{\widetilde{A}}(x_i)}{\sum\limits_{i=1}^{L} \overline{\mu}_{\widetilde{A}}(x_i) + \sum\limits_{i=L+1}^{n} \underline{\mu}_{\widetilde{A}}(x_i)},$$

$$y_R(\widetilde{A}) = \max_{r\in[1,n-1]} \frac{\sum\limits_{i=1}^{r} x_i\underline{\mu}_{\widetilde{A}}(x_i) + \sum\limits_{i=r+1}^{n} x_i\overline{\mu}_{\widetilde{A}}(x_i)}{\sum\limits_{i=1}^{r} \underline{\mu}_{\widetilde{A}}(x_i) + \sum\limits_{i=r+1}^{n} \overline{\mu}_{\widetilde{A}}(x_i)}$$

$$= \frac{\sum\limits_{i=1}^{R} x_i\underline{\mu}_{\widetilde{A}}(x_i) + \sum\limits_{i=R+1}^{n} x_i\overline{\mu}_{\widetilde{A}}(x_i)}{\sum\limits_{i=1}^{R} \underline{\mu}_{\widetilde{A}}(x_i) + \sum\limits_{i=R+1}^{n} \overline{\mu}_{\widetilde{A}}(x_i)}.$$

The KM algorithm [33] locates the switch points L, R and, as a consequence, the generalized centroid of $\widetilde{A}$ defined in Eq. (2.29), as explained in the following.

KM Algorithm to Calculate y_L
1. Calculate the initial point:

$$y' = \frac{\sum\limits_{i=1}^{n} x_i\theta_i}{\sum\limits_{i=1}^{n} \theta_i}, \quad \text{with} \quad \theta_i = \frac{\underline{\mu}_{\widetilde{A}}(x_i) + \overline{\mu}_{\widetilde{A}}(x_i)}{2}, \quad i = 1, 2, \ldots, n.$$

2. Find $1 \leq k \leq n - 1$ such that $x_k \leq y' \leq x_{k+1}$.
3. Define

$$\theta_i = \begin{cases} \overline{\mu}_{\tilde{A}}(x_i) \text{ if } i \leq k, \\ \underline{\mu}_{\tilde{A}}(x_i) \text{ if } i > k, \end{cases}$$

and calculate,

$$y_l(k) = \frac{\sum\limits_{i=1}^{n} x_i \theta_i}{\sum\limits_{i=1}^{n} \theta_i}.$$

4. If $y_l(k) = y'$, then stop and define $y_L = y_l(k)$, $L = k$. If not, go to step 5.
5. Define $y' = y_l(k)$ and go to step 2.

KM Algorithm to Calculate y_R
1. Calculate the initial point:

$$y' = \frac{\sum\limits_{i=1}^{n} x_i \theta_i}{\sum\limits_{i=1}^{n} \theta_i}, \quad \text{with} \quad \theta_i = \frac{\underline{\mu}_{\tilde{A}}(x_i) + \overline{\mu}_{\tilde{A}}(x_i)}{2}, \quad i = 1, 2, \ldots, n.$$

2. Find $1 \leq k \leq N - 1$ such that $x_k \leq y' \leq x_{k+1}$.
3. Define

$$\theta_i = \begin{cases} \underline{\mu}_{\tilde{A}}(x_i) \text{ if } i \leq k, \\ \overline{\mu}_{\tilde{A}}(x_i) \text{ if } i > k, \end{cases}$$

and calculate,

$$y_r(k) = \frac{\sum\limits_{i=1}^{n} x_i \theta_i}{\sum\limits_{i=1}^{n} \theta_i}.$$

4. If $y_r(k) = y'$, then stop and define $y_R = y_r(k)$, $R = k$. If not, go to step 5.
5. Define $y' = y_r(k)$ and go to step 2.

The goal of the Example 2.13 that was proposed by Castillo [9] and Castillo et al. [10], is to provide a better understanding of the Mamdani inference method for interval type-2 fuzzy sets and the KM algorithm.

Example 2.13 Consider the interval type-2 FRBS where the input variable is the temperature varying in the interval [0, 38] and the output variable is the thermal sensation, measured in the interval [0, 1]. The linguistic terms of the input variable

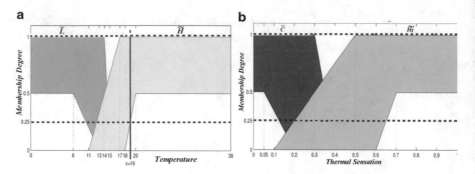

Fig. 2.31 FOUs of temperature and thermal sensation. (**a**) FOU of the input. (**b**) FOU of the output

are Low ($\widetilde{L}$) and High ($\widetilde{H}$) as shown in Fig. 2.31a. The linguistic terms of the output variable are Cold ($\widetilde{C}$) and Hit ($\widetilde{H}t$) shown in Fig. 2.31b.

The fuzzy rules are:

R^1 : If temperature is $\widetilde{L}$, then thermal sensation is $\widetilde{C}$;
R^2 : If temperature is $\widetilde{H}$, then thermal sensation is $\widetilde{H}t$.

The upper and lower membership functions of the input and output variables are defined as

$$\overline{\mu}_{\widetilde{L}}(x) = \begin{cases} 1 & \text{if } 0 \leq x < 14, \\ 15 - x & \text{if } 14 \leq x < 15, \\ 0 & \text{otherwise,} \end{cases} \quad \underline{\mu}_{\widetilde{L}}(x) = \begin{cases} 0.5 & \text{if } 0 \leq x < 8, \\ 1.3 - 0.1x & \text{if } 8 \leq x < 13, \\ 0 & \text{otherwise.} \end{cases}$$

$$\overline{\mu}_{\widetilde{H}}(x) = \begin{cases} \frac{x-11}{6} & \text{if } 11 \leq x < 17, \\ 1 & \text{if } 17 \leq x \leq 38, \\ 0 & \text{otherwise,} \end{cases} \quad \underline{\mu}_{\widetilde{H}}(x) = \begin{cases} 0.25(x - 18) & \text{if } 18 \leq x < 20 \\ 0.5 & \text{if } 20 \leq x \leq 38, \\ 0 & \text{otherwise,} \end{cases}$$

$$\overline{\mu}_{\widetilde{C}}(x) = \begin{cases} 1 & \text{if } 0 \leq x < 0.3, \\ 4 - 10x & \text{if } 0.3 \leq x < 0.4, \\ 0 & \text{otherwise,} \end{cases} \quad \underline{\mu}_{\widetilde{C}}(x) = \begin{cases} 0.5 & \text{if } 0 \leq x < 0.05, \\ \frac{0.1-0.5x}{0.15} & \text{if } 0.05 \leq x < 0.2, \\ 0 & \text{otherwise.} \end{cases}$$

$$\overline{\mu}_{\widetilde{H}t}(x) = \begin{cases} 2.5(x - 0.1) & \text{if } 0.1 \leq x < 0.5, \\ 1 & \text{if } 0.5 \leq x \leq 1, \\ 0 & \text{otherwise,} \end{cases} \quad \underline{\mu}_{\widetilde{H}t}(x) = \begin{cases} 5x - 3 & \text{if } 0.6 \leq x < 0.7, \\ 0.5 & \text{if } 0.7 \leq x \leq 1, \\ 0 & \text{otherwise.} \end{cases}$$

Let $x' = 19\,°\text{C}$ be the input point to be calculated, shown in Fig. 2.31a. Then, $\underline{\mu}_{\widetilde{H}}(19) = 0.25$ and $\overline{\mu}_{\widetilde{H}}(19) = 1$, thus $I_{x'} = [0.25, 1]$. The consequent of the thermal sensation is $\widetilde{H}t$ and the activation interval is $[0.25, 1]$, thus using

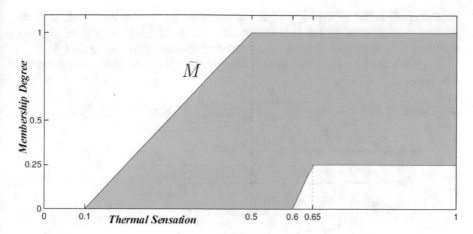

Fig. 2.32 The set $\widetilde{M}$

Table 2.6 Lower and upper membership degrees for the discrete set Q

i	x_i	$\underline{\mu}_{\widetilde{M}}(x_i)$	$\overline{\mu}_{\widetilde{M}}(x_i)$
1	0.1	0	0
2	0.2	0	0.25
3	0.3	0	0.5
4	0.4	0	0.75
5	0.5	0	1
6	0.6	0	1
7	0.7	0.25	1
8	0.8	0.25	1
9	0.9	0.25	1
10	1	0.25	1

the $t-$norm of the **minimum** (see Example 2.1), it is determined the output set $\widetilde{M}$, shown in Fig. 2.32. Notice this is the output of the interval type-2 Mamdani inference method. To defuzzify $\widetilde{M}$, it is used the KM algorithm. Therefore, the upper and lower membership functions of $\widetilde{M}$ are given by

$$\overline{\mu}_{\widetilde{M}}(x) = \begin{cases} 2.5(x - 0.1) & \text{if } 0.1 \leq x < 0.5, \\ 1 & \text{if } 0.5 \leq x \leq 1, \\ 0 & \text{otherwise,} \end{cases}$$

$$\underline{\mu}_{\widetilde{M}}(x) = \begin{cases} 5x - 3 & \text{if } 0.6 \leq x < 0.65, \\ 0.25 & \text{if } 0.65 \leq x \leq 1, \\ 0 & \text{otherwise.} \end{cases}$$

The set $Q = \{x_i = 0.1 \cdot i, \ i = 1, 2, \ldots, 10\} \subset [0.1, \ 1]$ is the chosen discrete set of the calculation for the generalized centroid of set $\tilde{M}$. Shown in Table 2.6 is the lower and upper membership degrees of these points of the universe of $\tilde{M}$.

Applying the KM algorithm to calculate y_L, it is followed the steps previously described.

(a) Define, $\theta_i = \dfrac{\underline{\mu}_{\tilde{M}}(x_i) + \overline{\mu}_{\tilde{M}}(x_i)}{2}, \ i = 1, \ldots, 10$. Then,

$$\theta_1 = \frac{\underline{\mu}_{\tilde{M}}(x_1) + \overline{\mu}_{\tilde{M}}(x_1)}{2} = \frac{0 + 0}{2} = 0;$$

$$\theta_2 = \frac{\underline{\mu}_{\tilde{M}}(x_2) + \overline{\mu}_{\tilde{M}}(x_2)}{2} = \frac{0 + 0.25}{2} = 0.125;$$

$$\theta_3 = \frac{\underline{\mu}_{\tilde{M}}(x_3) + \overline{\mu}_{\tilde{M}}(x_3)}{2} = \frac{0 + 0.5}{2} = 0.25;$$

$$\theta_4 = \frac{\underline{\mu}_{\tilde{M}}(x_4) + \overline{\mu}_{\tilde{M}}(x_4)}{2} = \frac{0 + 0.75}{2} = 0.375;$$

$$\theta_5 = \frac{\underline{\mu}_{\tilde{M}}(x_5) + \overline{\mu}_{\tilde{M}}(x_5)}{2} = \frac{0 + 1}{2} = 0.5;$$

$$\theta_6 = \frac{\underline{\mu}_{\tilde{M}}(x_6) + \overline{\mu}_{\tilde{M}}(x_6)}{2} = \frac{0 + 1}{2} = 0.5;$$

$$\theta_7 = \frac{\underline{\mu}_{\tilde{M}}(x_7) + \overline{\mu}_{\tilde{M}}(x_7)}{2} = \frac{0.25 + 1}{2} = 0.625;$$

$$\theta_8 = \frac{\underline{\mu}_{\tilde{M}}(x_8) + \overline{\mu}_{\tilde{M}}(x_8)}{2} = \frac{0.25 + 1}{2} = 0.625;$$

$$\theta_9 = \frac{\underline{\mu}_{\tilde{M}}(x_9) + \overline{\mu}_{\tilde{M}}(x_9)}{2} = \frac{0.25 + 1}{2} = 0.625;$$

$$\theta_{10} = \frac{\underline{\mu}_{\tilde{M}}(x_{10}) + \overline{\mu}_{\tilde{M}}(x_{10})}{2} = \frac{0.25 + 1}{2} = 0.625.$$

(b) Calculate the initial point, $y' = \dfrac{\displaystyle\sum_{i=1}^{10} x_i \theta_i}{\displaystyle\sum_{i=1}^{10} \theta_i}$, that is,

$$\sum_{i=1}^{10} x_i\theta_i = 0.1(0) + 0.2(0.125) + 0.3(0.25) + 0.4(0.375) + 0.5(0.5) + 0.6(0.5)$$

$$+ 0.7(0.625) + 0.9(0.625) + 1(0.625) = 2.925,$$

and $\sum_{i=1}^{10} \theta_i = 4.25$. Then, $y' = \dfrac{2.925}{4.25} = 0.688235294$.

(c) Find k, $1 \leq k \leq 9$, such that $x_k \leq y' \leq x_{k+1}$. As $x_6 < y' < x_7$, that is, $0.6 < y' < 0.7$, therefore $k = 6$.

(d) Define $\theta_i = \begin{cases} \overline{\mu}_{\widetilde{M}}(x_i) & \text{if } i \le 6, \\ \underline{\mu}_{\widetilde{M}}(x_i) & \text{if } i > 6, \end{cases}$ and calculate, $y_l(k) = \dfrac{\sum\limits_{i=1}^{10} x_i \theta_i}{\sum\limits_{i=1}^{10} \theta_i}$, that is,

$$y_l(k) = \frac{2.45}{4.5} = 0.544.$$

(e) As $y_l(k) \neq y'$ then continue the process, redefine $y' = y_l(k)$.

(f) Find k, $1 \le k \le 9$, such that $x_k \le y' \le x_{k+1}$. Let $x_5 < y' < x_6$, that is, $0.5 < y' < 0.6$, hence $k = 5$.

Define $\theta_i = \begin{cases} \overline{\mu}_{\widetilde{M}}(x_i) & \text{if } i \le 5, \\ \underline{\mu}_{\widetilde{M}}(x_i) & \text{if } i > 5, \end{cases}$ and calculate, $y' = \dfrac{\sum\limits_{i=1}^{10} x_i \theta_i}{\sum\limits_{i=1}^{10} \theta_i}$. Then,

$$y_l(k) = \frac{1.85}{3.5} = 0.528571428.$$

As $y_l(k) \neq y'$ then continue the process. Define $y' = y_l(k)$.

Find k, $1 \le k \le 9$, such that $x_k \le y' \le x_{k+1}$. For $k = 5$, these inequalities $x_5 < y' < x_6$ are true, that is, $0.5 < y' < 0.6$.

Define, $\theta_i = \begin{cases} \overline{\mu}_{\widetilde{M}}(x_i) & \text{if } i \le 5, \\ \underline{\mu}_{\widetilde{M}}(x_i) & \text{if } i > 5, \end{cases}$ and calculate, $y_l(k) = \dfrac{\sum\limits_{i=1}^{10} x_i \theta_i}{\sum\limits_{i=1}^{10} \theta_i}$. Then,

$$y_l(k) = \frac{1.85}{3.5} = 0.528571428.$$

As $y_l(k) = y'$, stop the process, being $L = 5$ and $y_L = 0.528571428$.

Now, applying the KM algorithm to compute y_R, the next steps of the algorithm are followed.

(a) Define, $\theta_i = \dfrac{\underline{\mu}_{\widetilde{M}}(x_i) + \overline{\mu}_{\widetilde{M}}(x_i)}{2}$, $i = 1, \ldots, 10$. The values θ_i, $i = 1, \ldots, 10$, were found previously.

(b) Calculate the initial point, $y' = \dfrac{\sum\limits_{i=1}^{10} x_i \theta_i}{\sum\limits_{i=1}^{10} \theta_i}$. This value was determined before,

$y' = 0.688235294.$

(c) Find k, $1 \le k \le 9$, such that $x_k \le y' \le x_{k+1}$. For $k = 6$, these inequalities $x_6 < y' < x_7$ are true, that is, $0.6 < y' < 0.7$.

(d) Define $\theta_i = \begin{cases} \underline{\mu}_{\widetilde{M}}(x_i) & \text{if } i \le 6, \\ \overline{\mu}_{\widetilde{M}}(x_i) & \text{if } i > 6, \end{cases}$ and calculate, $y_r(k) = \dfrac{\sum\limits_{i=1}^{10} x_i \theta_i}{\sum\limits_{i=1}^{10} \theta_i}$. Then,

$$y_r(k) = \frac{3.4}{4} = 0.85.$$

(e) As $y_r(k) \ne y'$ then continue the process. Define, $y' = y_r(k)$.

(f) Find k, $1 \le k \le 9$, such that $x_k \le y' \le x_{k+1}$. For $k = 8$, these inequalities $x_8 < y' < x_9$ are true, that is, $0.8 < y' < 0.9$.

Define $\theta_i = \begin{cases} \underline{\mu}_{\widetilde{M}}(x_i) & \text{if } i \le 8, \\ \overline{\mu}_{\widetilde{M}}(x_i) & \text{if } i > 8, \end{cases}$ and compute, $y_r(k) = \dfrac{\sum\limits_{i=1}^{10} x_i \theta_i}{\sum\limits_{i=1}^{10} \theta_i}$, that is,

$$y_r(k) = \frac{2.275}{2.5} = 0.91.$$

As $y_r(k) \ne y'$, then continue the process. Define $y' = y_r(k)$.

Determine k, $1 \le k \le 9$, such that $x_k \le y' \le x_{k+1}$. For $k = 9$, these inequalities $x_9 < y' < x_{10}$ are true, that is, $0.9 < y' < 1$.

Define, $\theta_i = \begin{cases} \underline{\mu}_{\widetilde{M}}(x_i) & \text{if } i \le 9, \\ \overline{\mu}_{\widetilde{M}}(x_i) & \text{if } i > 9, \end{cases}$ and calculate, $y_r(k) = \dfrac{\sum\limits_{i=1}^{10} x_i \theta_i}{\sum\limits_{i=1}^{10} \theta_i}$. Then

$$y_r(k) = \frac{1.6}{1.75} = 0.914285714.$$

As $y_r(k) \ne y'$ then continue the process. Define $y' = y_r(k)$.

Find k, $1 \le k \le 9$, such that $x_k \le y' \le x_{k+1}$. For $k = 9$, these inequalities $x_9 < y' < x_{10}$ are true, that is, $0.9 < y' < 1$.

Define $\theta_i = \begin{cases} \underline{\mu}_{\widetilde{M}}(x_i) & \text{if } i \le 9, \\ \overline{\mu}_{\widetilde{M}}(x_i) & \text{if } i > 9, \end{cases}$ and calculate, $y_r(k) = \dfrac{\sum\limits_{i=1}^{10} x_i \theta_i}{\sum\limits_{i=1}^{10} \theta_i}$, that is,

$$y_r(k) = \frac{1.6}{1.75} = 0.914285714.$$

As $y_r(k) = y'$, stop the process. Thus, $y_R = 0.914285714$ and $R = 9$.

Therefore, the defuzzified value is given by

$$y' = \frac{y_L + y_R}{2} = \frac{0.528571428 + 0.914285714}{2} = 0.721428571.$$

As a conclusion, taking $x' = 19°C$ the thermal sensation is approximately 0.72.

Exercise 2.8 As in Example 2.13 determine the thermal sensation corresponding to the temperature $x' = 9\,°C$.

2.3 Summary

Various topics have been expounded in this chapter.

- After succinct review of type-1 fuzzy set theory, the p-fuzzy systems are examined from two different point of view: the Takagi–Sugeno–Kang or Mamdani inference method.
- A review of a fuzzy system identification technique that involves a clustering process combined with a Takagi–Sugeno–Kang inference method.
- A type-2 fuzzy set, $\widetilde{A}$, is characterized by its membership function $\mu_{\widetilde{A}}$ defined in $X \times [0, 1]$, $\mu_{\widetilde{A}}(x, u) \in [0, 1]$, where X is the universe of discourse.
- When $\mu_{\widetilde{A}}(x, u) = 1$ for all (x, u) of the domain of $\mu_{\widetilde{A}}$ then $\widetilde{A}$ is called interval type-2 fuzzy set.
- Note that the set $\widetilde{A}(x)$ is a type-1 fuzzy set whose membership function is called secondary membership of x and its support is denoted by I_x.
- The upper membership function of $\widetilde{A}$ evaluated at $x \in X$ and denoted by $\overline{\mu}_{\widetilde{A}}(x)$ is the supremum of the set I_x. Likewise, the lower membership function of $\widetilde{A}$ evaluated at $x \in X$ and denoted by $\underline{\mu}_{\widetilde{A}}(x)$ is the infimum of the set I_x.
- A type-1 fuzzy set, A_e is said to be embedded in the interval type-2 fuzzy set, $\widetilde{A}$, if its membership function $\mu_{A_e}(x)$ satisfies $\underline{\mu}_{\widetilde{A}}(x) \leq \mu_{A_e}(x) \leq \overline{\mu}_{\widetilde{A}}(x)$ for all $x \in X$.
- The generalized (discrete) centroid of an interval type-2 fuzzy set $\widetilde{A}$ is the average between min $E_{\widetilde{A}}$ and max $E_{\widetilde{A}}$, where $E_{\widetilde{A}}$ is the set of Definition 2.16. To determine these values, the Karnik–Mendel algorithm is used.
- Examples and exercises are proposed to illustrate the interval type-2 fuzzy rule-based systems, utilizing diverse kind of inference methods.

References

1. Abonyi, J., Babuska, R., Szeifert, F.: Modified Gath-Geva fuzzy clustering for identification of Takagi-Sugeno fuzzy models. IEEE Trans. Syst. Man Cybern. **5**(1), 612–621 (2002)
2. Babuska, R.: Fuzzy Algorithms for Control, 2nd edn. Kluwer Academic, Delft

 3. Barros, L.C., Bassanezi, R.C.: Tópicos de Lógica Fuzzy e Biomatemática (in Portuguese), vol. 5, 2nd edn. Coleção MECC - Textos Didáticos, Campinas, Brazil (2010)
 4. Barros, L.C., Bassanezi, R.C., Lodwick, W.A.: A First Course in Fuzzy Logic, Fuzzy Dynamical Systems, and Biomathematics. Studies in Fuzziness And Soft Computing, vol. 347. Springer, Berlin (2017)
 5. Bassanezi, R.C.: Ensino-Aprendizagem com Modelagem Matemática (in Portuguese). Editora Contexto, Brazil (2002)
 6. Bezdek, J.: Pattern Recognition with Fuzzy Objective Function Algorithms. Plenum Press, New York (1981)
 7. Burden, R.L., Faires, J.D.: Numerical Analysis, 9th edn. Brooks/Cole Cengage Learning, San Francisco (2011)
 8. Cabrera, N.V.: Aplicação da extensão de Zadeh para conjuntos fuzzy tipo 2 intervalar (in Portuguese). Master's Thesis, Universidade Federal de Uberlândia, Uberlândia - Brazil (2014)
 9. Castillo, E.R.: Modelagem da dinâmica de um grupo de indivíduos HIV positivos com parâmetro fuzzy do tipo 2 (in Portuguese). Master's Thesis, Universidade Federal de Uberlândia, Uberlândia - Brazil (2014)
10. Castillo, E.R., Jafelice, R.M.: Modelagem de Indivíduos HIV Positivos com Parâmetro Fuzzy do Tipo 2 - Um aporte na Biomatemática, 1st edn. Novas Edições Acadêmicas, Niemcy (in Portuguese) (2016)
11. Castro, J.R., Castillo, O., Martínez, L.G.: Interval type-2 fuzzy logic toolbox. Eng. Lett. **15**(1), 1–10 (2007)
12. Cauchy, A.: Méthode générale pour la résolution des systèmes déquations simultanées. C. R. Acad. Sci. **25**, 536–538 (1847)
13. Dunn, J.C.: Well-separated clusters and optimal fuzzy partitions. J. Cybern. **4**(1), 95–104 (1974)
14. Ferreira, D.P.L.: Sistema p-fuzzy aplicado às equações diferenciais parciais (in Portuguese). Master's Thesis, Universidade Federal de Uberlândia, Uberlândia - Brazil (2011)
15. Ferreira, D.P.L., Jafelice, R.S.M., Serqueira, E.O.: Using fuzzy system in the study of luminescence and potency of neodymium ions. Appl. Opt. **51**, 6745–6752 (2012)
16. Gomes, L.T., Barros, L.C., Bede, B.: Fuzzy Differential Equations in Various Approaches. Springer Briefs in Mathematics SBMAC, 1st edn. Springer, Berlin (2015)
17. Gustafson, D., Kessel, W.: Fuzzy clustering with a fuzzy covariance matrix. Proc. IEEE Control Decis. Conf. **1**(1), 761–766 (1979)
18. Hohenwarter, M.: Geogebra 5.0 (2020). Accessed Sept. 2020. http://www.geogebra.org
19. Jafelice, R.M., Barros, L.C., Bassanezi, R.C.: Teoria dos Conjuntos Fuzzy. Springer Briefs in Mathematics SBMAC (in Portuguese), vol. 17, 2nd edn. Springer, Berlin (2012)
20. Jafelice, R.M., Bertone, A.M., Bassanezi, R.C.: A study on subjectivities of type 1 and 2 in parameters of differential equations. Tendências em Matemática Aplicada e Computacional **16**(1), 51–60 (2015)
21. Jang, J.S.R.: Anfis: Adaptive-network-based fuzzy inference systems. IEEE Trans. Syst. Man Cybern. **23**(3), 665–685 (1993)
22. Jouan-Rimbaud, D., Massart, D., Maesschalck, R.J.D.: The mahalanobis distance. Chemom. Intell. Lab. Syst. **50**(1), 1–18 (2000)
23. Karnik, N.N., Mendel, J.M.: Centroid of a type−2 fuzzy set. Inf. Sci. **132**, 195–220 (2001)
24. Larsen, P.M.: Industrial applications of fuzzy logic control. Int. J. Man-Machine Stud. **12**, 3–10 (1980)
25. Legendre, A.M.: Nouvelles méthodes pour la détermination des orbites des comètes (in French). F. Didot - Comets, Paris (1805)
26. Lémarechal, C.: Cauchy and the gradient method. Doc. Math. Extra 251–254 (2012). http://emis.maths.adelaide.edu.au/journals/DMJDMV/vol-ismp/40_lemarechal-claude.pdf
27. Malthus, T.R.: An Essay on the Principle of Population. J. Johnson, London (1798)
28. Mamdani, E.H., Assilian, S.: An experiment in linguistic synthesis with a fuzzy logic controller. Int. J. Man-Machine Stud.**7**, 1–13 (1975)

29. Martins, J.B., Bertone, A.M.A., Yamanaka, K.: Novel fuzzy system identification: comparative study and application for data forecasting. IEEE Latin Am. Trans. **17**, 1793–1799 (2019)
30. Massad, E., Ortega, N.R.S., Barros, L.C., Struchiner, C.J.: Fuzzy Logic in Action: Applications in Epidemiology and Beyond. Studies in Fuzziness And Soft Computing, vol. 232. Springer, Berlin (2008)
31. Mendel, J.M.: Type-2 fuzzy sets: some questions and answers. In: IEEE Neural Networks Society Newsletter, pp. 10–13 (2003)
32. Mendel, J.M.: Uncertain Rule-Based Fuzzy Logic Systems: Introduction and New Directions, 2nd edn. Prentice-Hall, Upper Saddle River (2017)
33. Mendel, J.M., Wu, H.: New results about the centroid of an interval type-2 fuzzy set, including the centroid of a fuzzy granule. Inf. Sci. **177**, 360–377 (2007)
34. Mendel, J.M., Wu, D.: Perceptual Computing Aiding People in Making Subjective Judgments. IEEE Press Series on Computational Intelligence. IEEE Press, Piscataway (2010)
35. Mendel, J.M., Rajati, M.R., Sussner, P.: On clarifying some definitions and notations used for type-2 fuzzy sets as well as some recommended changes. Inf. Sci. **340–341**, 337–345 (2016)
36. Pedrycz, W., Gomide, F.: An Introduction to Fuzzy Sets: Analysis and Design. Massachusetts Institute of Technology, Cambridge (1998)
37. Peixoto, M.S.: Sistemas dinâmicos e controladores fuzzy: um estudo da dispersão da morte súbita dos citros em São Paulo (in Portuguese). Ph.D. Thesis, UNICAMP, Campinas - Brazil (2005)
38. Peixoto, M.S., Barros, L.C., Bassanezi, R.C.: Predator - prey fuzzy model. Ecol. Model. **214**(1), 39–44 (2008)
39. RemarkAir$^{©:}$ https://remask.online/. RTKnits Ltd, Peupliers Avenue, Pointe aux Sables, Republic of Mauritius (2020)
40. Rizol, P.M.S.R., Mesquita, L., Saotome, O.: Lógica fuzzy tipo-2 (in Portuguese). Revista Sodebras **6**, 27–46 (2011)
41. Rumelhart, D.E., Hinton, G.E., Williams, R.J.: Learning representations by back-propagating errors. Nature **323**(6088), 533–536 (1986)
42. Stigler, S.M.: Gauss and the invention of least squares. Ann. Stat. **9**(3), 465–474 (1981)
43. Sugeno, M., Kang, G.T.: Structure identification on fuzzy model. Fuzzy Sets Syst. **28**, 329–346 (1988)
44. Takagi, T., Sugeno, M.: Fuzzy identification of systems and its applications to modeling and control. IEEE Trans. Syst. Man Cybern. **15**(1), 116–132 (1985)
45. The Mathwork, I.: Fuzzy Logic Toolbox - User's guide, 2018a edn., Natick (2018)
46. Verhulst, P.F.: Notice sur la loi que la population suit dans son accroissement. Correspondance mathématique et physique **10**, 113–121 (1838)
47. Website, W.H.O.: https://www.who.int/. Accessed July 2020
48. WHO: Multicentre growth reference study. Tech. Rep., World Health Organization (2003). https://www.who.int/childgrowth/mgrs/en/
49. Wu, D.: A brief tutorial on interval type-2 fuzzy sets and systems. http://www.learningace.com/doc/782209/2a080752ce7c48a761b8d3fa766db413/a-brief-tutorial-on-interval-type-2-fuzzy-sets-and-systems. Accessed: May, 2013
50. Zadeh, L.A.: Fuzzy sets. Inf. Control **8**, 338–353 (1965)
51. Zadeh, L.A.: The concept of a linguistic variable and its application to approximate reasoning-1. Inf. Sci. **8**, 199–249 (1975)

Chapter 3
Interval Type-2 Fuzzy Rule-Based System Applications

The finest minds seem to be formed rather by efforts at original thinking, by endeavors to form new combinations, and to discover new truths, than by passively receiving the impressions of other men's ideas [32].

Thomas Malthus (1798)

3.1 Elimination of Drugs: Interval Type-2 Fuzzy Sets Modeling the Uncertainty of Pharmacokinetics Action

Drugs are absorbed, distributed, and eliminated through different paths, according to their physical and chemical features. The main mechanism for drug excretion is the renal system. The classical models proposed for elimination of drugs are, in general, artificial and incomplete, in order to represent the complexity of the organisms in the elimination of drugs [50]. In [24, 31] a model using type-1 FRBS has been proposed with the aim of better representing the metabolic processes of the drug. The main part of this process is the velocity of the elimination of drugs from the organism, requiring an expert knowledge from the medical area. The velocity of elimination is uncertain and it is strongly dependent on the renal function. In this methodology, the rate of drug elimination is a parameter of the ordinary differential equation that models the phenomenon, and dependents on urinary volume, creatinine clearance, and blood pH values. The cases of three individuals are analyzed, based on medical information, assuming specific values for urinary volume, creatinine clearance, and blood pH values. From this information, an interval type-2 FRBS is built for modeling the velocity of the elimination of drugs. The same three sampled individuals are used to obtain a range for the decay of the concentration of drugs as a function of time. In two of the cases the drug saturation level is above that the recommended by pharmacokinetics and, as a consequence, the amount of the drug dose is decreased. Therefore, it appears that this procedure results in a satisfactory decision for the treatment of individuals.

R. S. da Motta Jafelice, A. M. A. Bertone, *Biological Models via Interval Type-2 Fuzzy Sets*, SpringerBriefs in Mathematics, https://doi.org/10.1007/978-3-030-64530-4_3

3.1.1   Classic Pharmacokinetics Model

Briefly reviewing the deterministic pharmacokinetics approach, the simplest model to describe the elimination of drugs is obtained when the rate of concentration variation of a drug is proportional to the drug concentration, $C = C(t)$, in the bloodstream [8], which is given by

$$\frac{dC}{dt} = -KC, \tag{3.1}$$

where K is the constant of the velocity of drug elimination [40]. An initial dose C_0 is given to an individual, and it is instantaneously absorbed into the bloodstream at instant $t = 0$. Thus, the solution of Eq. (3.1) is given by

$$C = C_0\exp(-Kt). \tag{3.2}$$

Define C_s as the maximum drug concentration acceptable by the individual organism. When a set of doses is given in equally spaced time-intervals (T), C_s has the following expression:

$$C_s = \frac{C_0}{1 - \exp(-KT)}, \tag{3.3}$$

which represents the drug saturation level for the individual considered. Represented in Fig. 3.1 is the concentration curve of a drug in the organism after four given doses.

The time that the plasmatic concentration of a specific drug needs to be reduced to its half is denominated as half-life time. Knowing this value is useful to achieve the maximum concentration of constant medium plasmatic after repeated dose-intervals.

From the solution of Eq. (3.1), if $t = t_{\frac{1}{2}}$, then $C = \dfrac{C_0}{2}$, it is possible to determine the relation between the half-life time and velocity of elimination of a drug, given by: $K = \frac{0.693}{t_{\frac{1}{2}}}$. Considering 1 h as the shortest half-life time, the value of K is 0.693, which is the one used in the model.

3.1.2   Pharmacokinetics Model Through Type-1 Fuzzy Sets

The type-1 fuzzy model developed in [31] differs from the deterministic approach Eq. (3.1) since K depends on the urinary volume, uv, the creatinine clearance, $crcl$, and the blood pH, p.

Through the knowledge of a medical expert in this area of study, the collected information about the velocity of elimination of the drugs depends on the following variables:

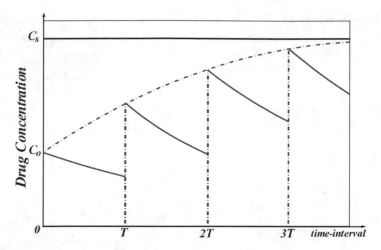

Fig. 3.1 Representation of the drug concentration after four given doses [8, 31]

Table 3.1 Quantities of urinary volume

Urinary volume	Quantity (ml)
Anuria (A)	[0, 100]
Oliguria (O)	[100, 300]
Normal (N)	[300, 1500]
Polyuria (P)	[1500, 3000]

Table 3.2 Classification for the creatinine clearance

crcl	Blood volume (ml/min)
Very low (VL)	[0, 10]
Low (L)	[10, 50]
Medium low (ML)	[50, 90]
Normal (N)	[90, 120]
High (H)	[120, 200]

1. Urinary volume: the urine production of an individual every 24 h, which is modeled according to the quantities shown in Table 3.1.
2. Creatinine Clearance: The creatinine clearance test determines the efficiency of the kidneys to eliminate the creatinine from the blood, which is a final-metabolism product from creatine. Creatine is produced in the liver and it provides energy for the muscles. It is in the blood-serum in the proportional amount to the muscular body mass. The clearance rate is expressed in terms of blood volume that can be free of creatinine in just 1 min. Creatinine clearance is classified depending on the blood volume per minute, using the classification shown in Table 3.2.
3. Blood pH: classified by level, it has been considered in the quantities shown in Table 3.3.

Table 3.3 Classification for the blood pH

Blood pH	Quantity
Acid (Ac)	[6.5, 7.35]
Normal (N)	[7.35, 7.45]
Basic (B)	[7.45, 8]

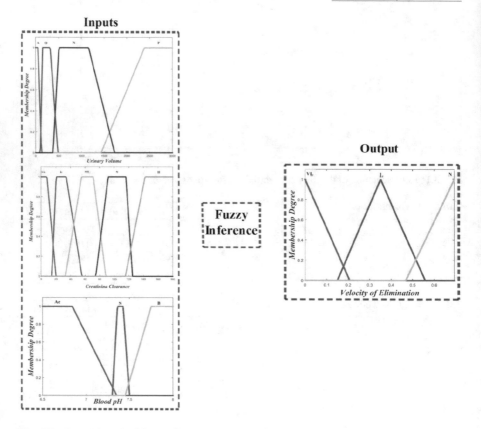

Fig. 3.2 Type-1 FRBS of the pharmacokinetics model

4. Velocity of elimination: defined in the interval [0, 0.693] has as linguistic terms Very Low (VL), Low (L), and Normal (N).

Based on this set of valuable information, a type-1 FRBS is built as depicted in Fig. 3.2. An expert in the area has helped to encode the relationship between uv, $crcl$, p, and K summarized in the rule base described in Table 3.4 for urinary volumes classified as anuria and oliguria; in Table 3.5 classified as normal and polyuria.

Given the type-1 FRBS detailed in Fig. 3.2 and using Mamdani inference method with center of gravity defuzzification technique, it is determined that $K = K(uv, crcl, p)$. From this model it is possible to obtain the concentration of drugs

Table 3.4 Fuzzy rules for uv when is anuria or oliguria

p	Ac	N	B	p	Ac	N	B
$crcl$				$crcl$			
VL	VL	VL	VL	VL	VL	VL	VL
L	VL	L	L	L	VL	L	L
ML	VL	L	L	ML	L	N	N
N	N	N	N	N	L	N	N
H	N	N	N	H	N	N	N
Anuria				Oliguria			

Table 3.5 Fuzzy rules for uv when is normal or polyuria

p	Ac	N	B	p	Ac	N	B
$crcl$				$crcl$			
VL	VL	VL	VL	VL	VL	VL	VL
L	N	L	L	L	L	L	L
ML	N	N	N	ML	N	N	N
N	N	N	N	N	N	N	N
H	N	N	N	H	N	N	N
Normal				Polyuria			

Table 3.6 Velocity of elimination of the drug for each individual via type-1 FRBS

Individual	uv	$crcl$	p	K
I_1	1500 ml daily	100 ml/min	7.4	0.6032
I_2	100 ml daily	10 ml/min	7.35	0.0860
I_3	300 ml daily	35 ml/min	7.25	0.2308

in each individual and provide information to the expert regarding individuals who present renal problems.

3.1.2.1 Case Study: Renal Insufficiency and the Elimination of Drugs Using Type-1 Fuzzy Set

When an individual takes any drug in 8 h interval, part of the drug will be absorbed and part eliminated by the body. The type-1 FRBS has been utilized in [31] to analyze how the urinary volume, uv, creatinine clearance, $crcl$, and blood pH, p influence in the velocity of the elimination of the drug (K).

The case study is about the prescription of 500 mg of a certain drug, in 8 h interval for three chosen individuals I_1, I_2, and I_3. The velocity of elimination of the drug for each individual through type-1 fuzzy model is shown in Table 3.6.

From the information of Table 3.6, it is possible to obtain the saturation level and elimination of the drugs for an individual through Eqs. (3.2) and (3.3) using the defuzzified value of $K = K(uv, clcr, p)$ resulting from the type-1 FRBS. The saturation level of I_1 is normal, 504 mg, the I_2, has saturation level around 1005 mg, which means that his renal function is at risk. The I_3 has his renal function at risk

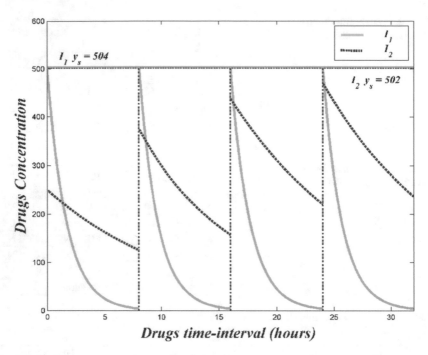

Fig. 3.3 Drugs concentration for individuals I_1 and I_2 with change of the prescriptive dose in the type-1 fuzzy model

although his saturation level is lower, around 594 mg. In order to avoid any risk of medicamentous intoxication, it is necessary to change the prescription for I_2 and I_3. Graphs with the complete analysis are detailed in [31].

In what follows is presented the I_2 analysis, being the worst of the two cases. The study of renal insufficiency, corresponding to the individual I_3, is carried out as in the case of I_2. Computationally, there has been established two alternatives for I_2 to decrease the saturation level:

1. time-interval between doses of the drugs has been changed to every 24 h. This has resulted in a decrease of the saturation level to 572 mg, which is close to an individual whose renal function is normal, nevertheless not enough;
2. change the prescribed dose from 500 mg to 250 mg and keeping the same time-interval, showing a decrease of the saturation level to normal level of 502 mg, as depicted in Fig. 3.3.

According to the expert this result is consistent with what happens with real patients whose renal function are most at risk and it indicates a necessary change in the drug's prescription.

The velocity of drug elimination can be obtained from each individual, according to Zanini [50]. Therefore, considering the velocity of drug elimination as a fuzzy parameter, it is possible to evaluate the renal function of an individual.

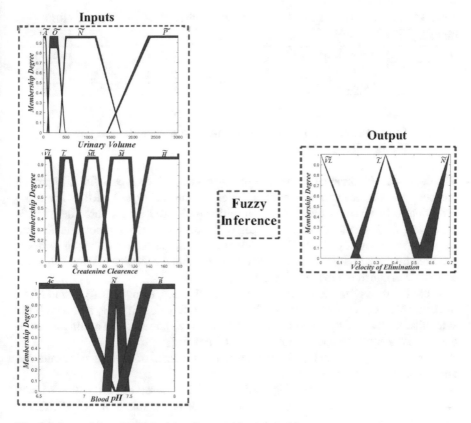

Fig. 3.4 Interval type-2 FRBS of the pharmacokinetics model

3.1.3 Pharmacokinetics Model Through Interval Type-2 Fuzzy Sets

Inspired by the drug concentration model with the elimination speed, K, obtained through a type-1 FRBS and in order to incorporate the different types of uncertainties of this phenomenon, an interval type-2 FRBS is built. As a result of this transformation of the model, y_L, y', and y_R are obtained corresponding to the data of each individual in Table 3.6. Therefore, Cs_L, Cs, and Cs_R are the saturation values of the drug concentration. The interval type-2 fuzzy rule base is the same as used in the type-1 fuzzy rule base, shown in Tables 3.4 and 3.5, substituting the name of the type-1 fuzzy set, that represents the linguistic term, for the corresponding name of the interval type-2 fuzzy set. For instance, the set L is transformed in $\widetilde{L}$, as depicted in Fig. 3.4. In the same figure is described the interval type-2 FRBS.

Mamdani inference method is also used for the interval type-2 FRBS and the generalized centroid as a defuzzification technique (see Sect. 2.2.1.3).

Table 3.7 Velocity of elimination of the drug for each individual via interval type-2 FRBS

Individual	uv	$crcl$	p	y_L	y'	y_R
I_1	1500 ml daily	100 ml/min	7.4	0.5128	0.6029	0.6930
I_2	100 ml daily	10 ml/min	7.35	0.0727	0.086	0.0992
I_3	300 ml daily	35 ml/min	7.25	0.0865	0.2307	0.3748

3.1.3.1 Case Study: Renal Insufficiency and the Elimination of Drugs Using Interval Type-2 Fuzzy Sets

The data from the three individuals in the case study of the type-1 fuzzy approach has been used to develop the analysis under the interval type-2 FRBS. The same prescription of 500 mg of a drug, in 8 h intervals, is used for the three individuals I_1, I_2, and I_3. The velocity of elimination of the drug for each individual is denoted by y', as shown in Eq. (2.29). This value along with the output values y_L and y_R, acquired by the KM algorithm , are presented in Table 3.7. Notice that the values y_L and y_R determined the range of uncertainty.

From the data of the three last columns of Table 3.7, using Eqs. (3.2) and (3.3), the concentration level of drugs is calculated as a function of time, along with the saturation level of each individual. The graphs in Fig. 3.5 represent a comparison of the drug concentration and saturation level of I_1 and I_2.

Similarly, the graphs in Fig. 3.6 represent a comparison of the drug concentration and saturation level between I_1 and I_3.

In the comparisons shown in Figs. 3.5 and 3.6, it is observed that, with the range of uncertainty, very common in biological phenomena, the values of the individuals' saturation levels have become higher. The uncertainty of the saturation level ranges are: [501, 508], [912, 1133], and [526, 1000], respectively, for I_1, I_2, and I_3.

Similar computational alternatives, as developed for type-1 FRBS, have been created for I_2: time-interval between doses of the drugs has been changed from each 8 h to 24 h; and a change in the prescribed dose from 500 mg to 250 mg, keeping the same time-interval. The result of the comparison between I_1 with I_2 is shown in Fig. 3.7, with an increase in time between doses of the drugs.

Compared in Fig. 3.8 are I_1 and I_2 when the prescription of the dose is decreased.

It can be seen in Fig. 3.7 that, with the time-interval change to 24 h for I_2, the saturation level uncertainty interval is [550, 605], still at risk of intoxication medication. Furthermore, in Fig. 3.8, notice that the dose prescribed for I_2 changed from 500 mg to 250 mg, keeping the time-interval between doses. In this case, the saturation interval is [456, 566]. The ranges of the two alternatives are above the saturation level. As the second alternative is in accordance with the medical protocols, the implementation of this alternative has been performed, decreasing the described dose to 220 mg. This has achieved the saturation level interval [401, 498] which is considered satisfactory. This result is shown in Fig. 3.9.

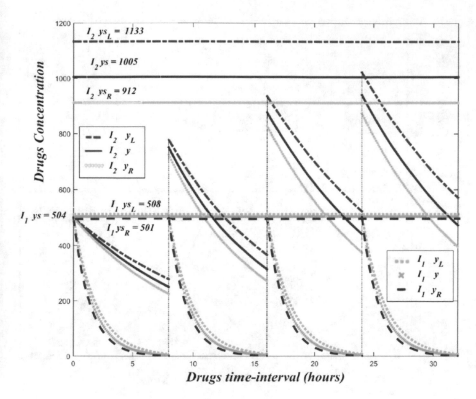

Fig. 3.5 Drugs concentration for individuals I_1 and I_2 in the interval type-2 fuzzy model

3.1.4 Conclusion

In fuzzy modeling with only some data from individuals with renal impairment and the knowledge of a nephrologist physician, it has been possible to construct an interval type-2 FRBS, which takes into account the vagueness of various types. This methodology can provide a range of alternatives for the speed of elimination of drugs for other data from individuals with renal impairment. The behavior of the downward curve of drug concentration into the bloodstream of each individual has verified a satisfactory result when the dose prescribed is diminished. The information provided by this mathematical modeling might be used in conjunction with the medical expert to avoid any risk of medicamentous intoxication.

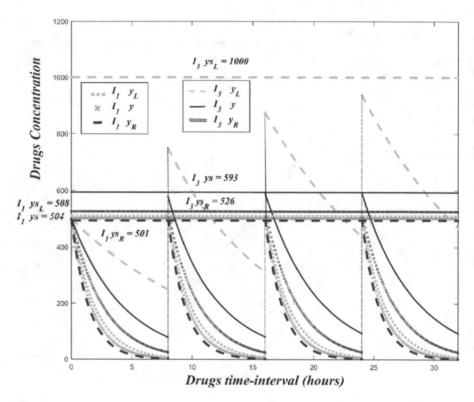

Fig. 3.6 Drug concentration for individuals I_1 and I_3 in the interval type-2 fuzzy model

3.2 Interval Type-2 Fuzzy Rule-Based System to Predict the Prostate Cancer Stage

Establishing the stage of prostate cancer is important to help the physicians define the therapy to be used. The purpose of this application is to verify that an interval type-2 FRBS can present a wider range of possibilities to identify patients with organ-confined prostate cancer, comparing solely with the output obtained through the type-1 fuzzy approach. This study has been developed by Castanho [10] with the aim to construct a system of classified patients in a database according to their cancer stage. Applying an interval type-2 fuzzy approach, it is obtained the so-called area under the Receiver Operating Characteristic [19], the measure to evaluate the system, are values in the interval [0.816, 0.837] that includes the type-1 FRBS output equal to 0.824. Notice that the value 0.837, being the closest to 1, is the one chosen to represent the result of the system, derived from embedded type-1 fuzzy sets in the input variables of the interval type-2 FRBS [13].

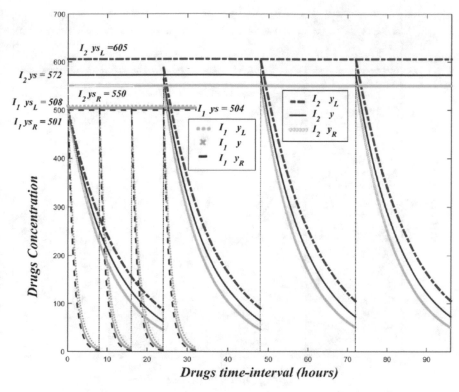

Fig. 3.7 Drug concentration for individuals I_1 and I_2 with alteration in the dose's time-interval in the interval type-2 fuzzy model

3.2.1   A Predict Prostate Cancer Stage Model Using Type-1 FRBS

According to the World Health Organization (WHO), prostate cancer is the second most common cancer among men, claiming half a million lives each year worldwide. When a man is diagnosed with localized prostate cancer, the probability that it is cured after surgery is approximately 80%. In case there is an extension beyond the prostate, it is important to discuss not only the cure but also the quality of life.

There are many studies to help the physicians establish the prognosis. Kattan et al. [28] developed nomograms to predict the probability that a man with prostate cancer has an indolent tumor based on clinical variables: Prostatic Specific Antigen (PSA), clinical stage, prostate biopsy Gleason grade, ultrasound volume, and other variables derived from the analysis of systematic biopsies. Steyerberg et al. [46] validated and updated a prognostic nomogram that predicts indolent prostate cancer. Indolent cancer is so small, low grade, and noninvasive that it offers little risk to the patient.

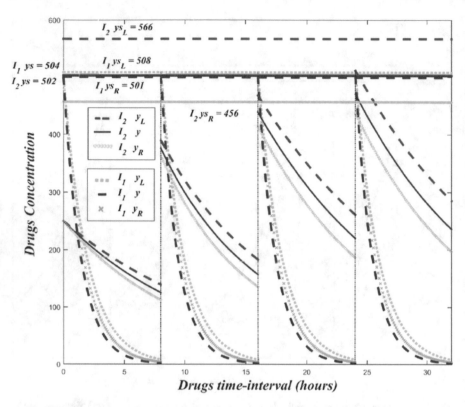

Fig. 3.8 Drugs concentration for individuals I_1 and I_2 with alteration in the prescription's dose in the interval type-2 fuzzy model

Soft computing techniques have been used for this purpose. Ren-Jieh et al. [41] developed a two-stage Fuzzy Neural Network using a Particle Swarm Optimization algorithm and concluded that if the clinical data are known, the prognosis of prostate cancer patients can be forecast. Ozkan et al. [36] utilized a type-2 fuzzy expert system to determine prostate cancer diagnosis. The analysis of the prostate cancer model is done through two points of view: type-1 and interval type-2 fuzzy approaches.

Castanho et al. [12] have used fuzzy set theory and Genetic Algorithms [21] to construct a fuzzy expert system to identify patients with confined or non-confined prostate cancer. Between January 1997 and June 2008, 331 patients ranging from 43 to 76 years old (median 64) were treated with radical prostatectomy for clinically localized prostate cancer at Clinics Hospital of State University of Campinas, Brazil. The surgical specimen (prostate and adjacent structures) of each patient was examined by the same pathologist and the tumor extension was confirmed: 289 patients had organ-confined cancer and 42 had cancer evident outside the prostatic capsule. The clinical stage is determined by Digital Rectal Examination (DRE) that depends on the perception and expertise of the physician; the tumor marker PSA

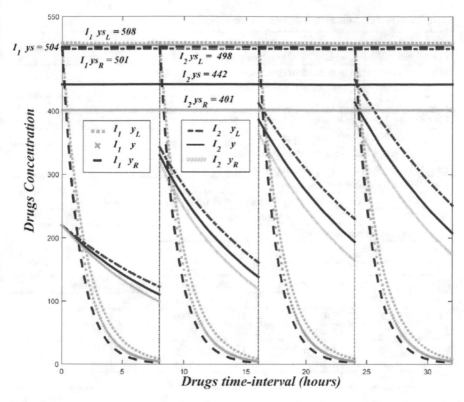

Fig. 3.9 Drugs concentration for individuals I_1 and I_2 with the prescription's dose equal to 220 mg in the interval type-2 fuzzy model

Table 3.8 PSA levels, where ng/mL stands for nanograms per milliliter

PSA (ng/mL)	Description
Until 4.0	Normal
4.0 to 10.0	Slightly elevated
10.0 to 20.0	Moderately elevated
Over 20.0	Highly elevated

level is related to pathological stage in Table 3.8, although higher preoperative serum PSA levels are not always associated with advanced pathological features, and lower values do not necessarily suggest organ-confined disease.

The biopsy result gives the Gleason score. This grade is assigned by a pathologist after the analysis of a tumor specimen and, in this way, a precise value is obtained that represents an imprecise situation.

The Clinical Stage labels, according to the TNM classification of Malignant Tumors, are a globally recognized standard for classifying the extent of the spread of prostate cancer. It is a classification system of the anatomical extent of tumor cancers in Table 3.9. The four clinical stages are: T1 (which includes T1a, T1b,

Table 3.9 TNM classification system of prostate cancer

Stage	Description
T1a	Non palpable, with ≤5% of tissue with cancer, low grade
T1b	Non palpable, with >5% of tissue with cancer and/or high grade
T1c	Non palpable, but prostate-specific antigen elevated
T2a	Palpable, half of 1 lobe or less
T2b	Palpable, more than half of 1 lobe, not both lobes
T2c	Palpable, involves both lobes
T3a	Palpable, unilateral capsular penetration
T3b	Palpable, bilateral capsular penetration
T3c	Positive seminal vesicle involvement
T4	Tumor is fixed or invades adjacent structures other than seminal vesicles
N	Positive pelvic lymph nodes involvement
M	Distant metastatic spread

Table 3.10 Classification of database patients

PSA	Patients	Gleason score	Patients	Clinical stage	Patients
0–2.5	15	4	3	T1a	9
2.6–4	23	5–6	211	T1b	4
4.1–6	72	7 = 3 + 4	84	T1c	149
6.1–8	63	7 = 4 + 3	17	T2a	118
8.1–10	49	8–10	16	T2b	41
>10	109			T2c	10

and T1c), T2a, T2b, and T2c. From stage T3a on, the tumor is beyond prostate and advanced stages are not considered in this model.

A classification of the disease's stage, extracted from the 311 patients analyzed, is shown in Table 3.10.

From the characteristics of the model, the variables are uncertain. To find a type-1 FRBS that classifies the patients into organ-confined cancer or non-confined, a genetic algorithm was developed [12]. The system obtained has the input variables are PSA level, Gleason score, and clinical stage.

The linguistic terms for the variable PSA level are: Normal (N), Slightly Elevated (SE), Moderately Elevated (ME), Very Elevated (VE), Highly Elevated (HE), or Extremely Elevated (EE).

The linguistic variable Gleason Score is classified with the following linguistic terms: Well-Differentiated (WD), tumor with less aggressive behavior; Moderately Differentiated (MD), and Poorly-Differentiated (PD) with more aggressive behavior.

For the output variable, Cancer Stage (y), the following linguistic terms are attributed: Confined (Co), if all cancer is confined within the prostatic capsule, and Non-Confined (NCo) if cancer is evidently outside the prostate. To facilitate the mathematical modeling of the applications proposed in this chapter, the fuzzy rules

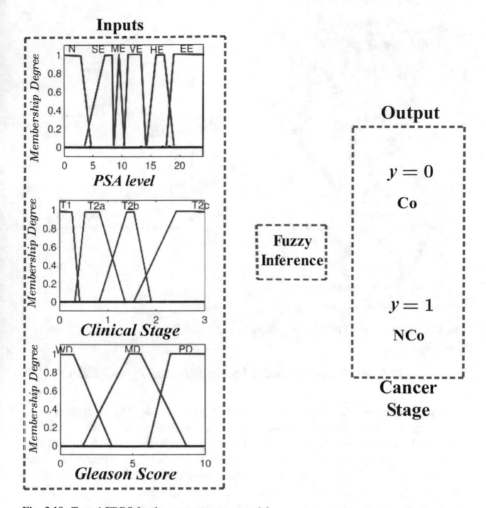

Fig. 3.10 Type-1 FRBS for the prostate cancer model

of TSK inference method are synthesized. For example, in this model, the output values are: $y = 0$ if Co and $y = 1$ if NCo. The rule base is composed by 47 rules like this:

"If PSA level is ME and Gleason Score is PD and Clinical Stage is T2b then $y = 1$."

The membership functions of inputs and output variables of the TSK inference method are shown in Fig. 3.10.

Mendel claims [33] that if a type-1 fuzzy set is used to model a phenomenon, then there also needs some measure of dispersion to capture more linguistic uncertainties. Inspired by this idea, from the type-1 FRBS an adapted interval type-2 FRBS is formulated. Thereby, the FRBS obtained include a measure of dispersion which allows flexibility [13].

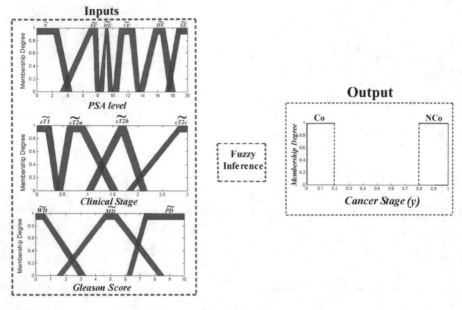

Fig. 3.11 Interval type-2 FRBS for the prostate cancer model

3.2.2   A Prostate Cancer Model Using Interval Type-2 FRBS

The interval type-2 fuzzy logic toolbox of Castro et al. [11] is applied as a computational mean for the construction of an interval type-2 FRBS based in the type-1 FRBS shown in Sect. 3.2.1. The only difference between the systems is that the input membership functions, shown in Fig. 3.10, are embedded in the interval type-2 fuzzy sets in Fig. 3.11, and the constant outputs for the type-1 FRBS become a type-1 interval fuzzy number. The same rule base of the type-1 FRBS is transformed to the interval type-2 FRBS substituting the name of the type-1 set, that represents the linguistic term, for the corresponding name of the interval type-2 fuzzy set. The components of the interval type-2 FRBS are presented in Fig. 3.11.

New data are tested with the interval type-2 FRBS constructed, shown in Fig. 3.11, using the previous data, in order to validate the system. In fact, between July 2008 and November 2011, 48 patients with average age of 63 were treated with radical prostatectomy for clinically same hospital in Campinas, Brazil. The surgical specimens (prostate and adjacent structures) from each patient were examined by the same pathologist aforesaid and the pathologic stage established: 34 patients had organ-confined cancer (TNM at level palpable T2) and 14 patients had extraprostatic cancer (TNM bigger than level palpable T2). It is the same database utilized to validate the type-1 FRBS [12].

Fig. 3.12 Venn's Diagram to illustrate the test result in a population

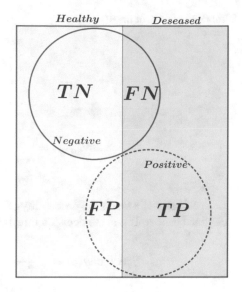

The Area Under the Curve, AUC, of the Receiver Operating Characteristic, ROC, is the measure to evaluate the system results, both type-1 and interval type-2 FRBS. The ROC principles are detailed in Sect. 3.2.3.

3.2.3 ROC Measure

A population that is submitted to a test has as a result x_i, $i = 1, 2, \ldots, n$, values corresponding to the $i - th$ individual. There is a threshold value, k, named the cut-off point that separates the population in two sets: Positive, P, and Negative N. If $x_i \geq k$, then result is Positive, P, otherwise the result is Negative, N.

For each cut-off point the outcomes of a test are depicted in Fig. 3.12, represented in the interior of the dashed circle is the Positive set and in the interior of the continuous line circle the Negative set, where:

TP is the true positive set, cases with the disease correctly classified as positive;
FP is the false positive set, cases without the disease classified as positive;
TN is the true negative set, cases without the disease correctly classified as negative;
FN is the false negative set, cases with the disease classified as negative.

Two important performance measures, *sensitivity* and *specificity*, can be calculated for each cut-off point. The test of sensitivity calculates the probability that the test correctly classifies a patient with non-confined cancer, depends on the choice of the cut-off point, k, which is the threshold decision. For the chosen one k, sensitivity is defined as

Table 3.11 AUC of the ROC curve

Results	AUC	95% CI
y_L	0.837	0.702 to 0.928
y'	0.831	0.695 to 0.923
y_R	0.816	0.678 to 0.913

$$\text{Sensitivity}(k) = \frac{\sum_{i=1}^{n} TP(x_i)}{\sum_{i=1}^{n} TP(x_i) + \sum_{i=1}^{n} FN(x_i).}$$

Specificity of a test is the probability of correctly classifying a patient with organ-confined cancer. If k is the decision threshold, one has

$$\text{Specificity}(k) = \frac{\sum_{i=1}^{n} TN(x_i)}{\sum_{i=1}^{n} TN(x_i) + \sum_{i=1}^{n} FP(x_i).}$$

A test is examined for its diagnostic performance by using a plot of the sensitivity on the y-axis and (1—specificity) on the x-axis. The curve resulting from continuously varying the decision threshold over the entire range of results observed is the ROC curve.

The AUC of the ROC is an important metric to evaluate the performance of a test. A test that achieves perfect diagnostic discrimination for a specific decision threshold has area AUC equal to 1 and when no apparent distributional difference between the two groups the area is 0.5. This area represents the probability that a randomly chosen diseased individual is ranked with greater suspicion than a randomly chosen non-diseased individual [19].

3.2.4 Comparison of the FRBS in ROC Measure

AUC of ROC has been used to evaluate the interval type-2 FRBS to develop the comparison. As seen in Sect. 3.2.3, this curve depends on two performance measures: sensitivity and specificity.

The output of the interval type-2 FRBS is given by y_L, y', and y_R in which $y' = \dfrac{y_L + y_R}{2}$ is the defuzzified value. Thus, three possible results are obtained for the 48 patients, which means one can calculate three ROC curves (see Table 3.11). The AUC shown in Fig. 3.13 is 0.837 (95%CI: 0.702 to 0.928). The cut-off point is $k = 0.5458$ for sensitivity equal to 0.7857 and specificity equal to 0.8824.

For the type-1 FRBS, the AUC of the ROC curve is 0.824, 95% CI: 0.686 to 0.918. The difference between the AUC of ROC curves is 0.013, showing that the interval type-2 methodology improves the type-1 fuzzy technique.

Fig. 3.13 ROC curve

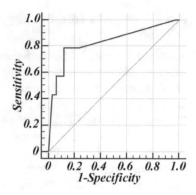

3.2.5   Conclusion

The use of an interval type-2 FRBS to predict the prostate cancer stage (organ-confined or non-confined) has shown an improvement over the type-1 fuzzy modeling. The evaluation of the system, made through the Area Under the Curve of the Receiver Operating Characteristic, in the case of the interval type-2 fuzzy approach, is an interval that includes the type-1 fuzzy result, widening the options to identify patients with organ-confined prostate cancer. The number that represents the closest value to 1 comes from the interval type-2 FRBS outcome, accomplishing the goal of this investigation.

3.3   Interval Type-2 p-Fuzzy System of the Transference from Asymptomatic to Symptomatic HIV-Seropositive Population

The aim of this work, based on Ferreira et al. [18], is to study the type-1 and interval type-2 p-fuzzy systems, defined in Sect. 2.1.3, to model the transference of Human Immunodeficiency Virus (HIV)-seropositive population from asymptomatic to symptomatic. The trajectories obtained by the type-1 and interval type-2 p-fuzzy systems are compared with a numerical approximation by integration of Peterman's data (1985), using the maximum absolute error at the same instant of the data. The interval type-2 p-fuzzy system's trajectory provides the best approximation. Furthermore, from the interval type-2 p-fuzzy system results a range containing the type-1 p-fuzzy system's trajectory and Peterman's data numerically integrated.

3.3.1 Human Immunodeficiency Virus (HIV)

Acquired Immunodeficiency Syndrome (AIDS) comes from an immunodeficiency process resulting from HIV infection. The disease compromises the immune system, which destroys the mechanisms of defense of the human organism, causing the loss of natural immunity and allowing the development of opportunistic diseases. The transmission of HIV occurs through sexual relations, blood transfusion, use of contaminated needles or surgical materials, breast-feeding, mucous contact, and perinatal transmission [35].

The first recorded information about the disease comes from Africa in 1920, and since then, according to AVERT organization [6], sporadic cases of AIDS were documented prior to 1970, suggesting that the current epidemic started in the mid- to late 1970s. By 1980, HIV may have already spread to five continents being recognized as an epidemic in the USA in 1981. In 1987 the first antiretroviral drug was approved and, since then, has enabled individuals to live long and healthy lives coping with HIV. Up to 2019, AIDS experts from the United Nations [47] have recorded about 38 million people worldwide who are infected with HIV. In this same year, the statistics shows that around 67% of all individuals living with HIV had access to antiretroviral drugs treatment [1].

Shown in Fig. 3.14 is the HIV cell structure, which is detailed in the following.

HIV is a spherical retrovirus, that is, a virus containing Ribonucleic Acid (RNA), that replicates in the $CD4+$ T lymphocyte host cell, being the main cells attacked by HIV when these reach the bloodstream. The HIV is encapsulated in a Gycoprotein Complex envelope. This protein wrap consists of three main proteins: a larger one, gp120, which forms buttons of Glycosylation on the surface, and a smaller one, gp41. In the internal capsule (capsid) there is a membrane that wraps the genetic material and the nucleocapsid associates with the genomic RNA. Two important proteins are inside the Capsid: Reverse Transcriptase and Integrase. The Protease is the third crucial protein.

The antiretroviral treatment inhibits the function of the three main virus proteins. These inhibitors prevent free virus particles from infecting $CD4+$ T lymphocytes and they delay the viral replication.

Described in Sect. 3.4 is a model that considered antiretroviral treatment.

3.3.2 Evolution Deterministic Model of Symptomatic HIV-Seropositive Population

The model presented in this section is the transfer from asymptomatic to symptomatic HIV-seropositive population. The behavioral model of converting from asymptomatic to symptomatic HIV-seropositive population, without treatment with antiretrovirals, has been studied by Anderson (1986)[3], who establishes that the rate of conversion from infection to AIDS is a function of time. This model is given by

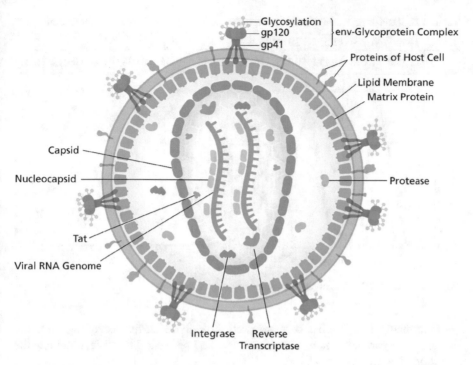

Fig. 3.14 The HIV structure. Image source: Thomas Splettstoesser, www.scistyle.com [45]

$$\frac{da}{dt} = -\lambda(t)a \qquad\qquad a(0) = 1 \qquad\qquad (3.4)$$

$$\frac{ds}{dt} = \lambda(t)a = \lambda(t)(1 - s) \qquad\qquad s(0) = 0, \qquad\qquad (3.5)$$

where $a + s = 1$, a represents the fraction of asymptomatic and s the fraction of symptomatic individuals, as a function of time. The parameter $\lambda(t)$ is the transference rate between infected individuals and infected individuals that develop AIDS, considered as a linear function of time $\lambda(t) = mt$, where m is a positive constant. Hence, Eq. (3.4) becomes $\frac{da}{dt} = -mta$. Solving this equation one obtains the solution as:

$$a(t) = \exp(-\frac{mt^2}{2}) \qquad\qquad s(t) = 1 - \exp(-\frac{mt^2}{2}). \qquad (3.6)$$

Peterman et al. [39] present data related with 194 cases of blood transfusion-associated AIDS, where approximated values of t_i, $i = 1, \ldots, 11$ and $\frac{ds}{dt}(t_i)$ are shown in Table 3.12.

Table 3.12 Data of Peterman approximation for the t_i, $i = 1, \ldots, 11$, years after infection

t_i	0.6	1.2	1.5	2.1	2.6	3.1	3.6	4	4.5	5	5.5
$\dfrac{ds}{dt}(t_i)$	0.048	0.078	0.172	0.115	0.155	0.088	0.155	0.038	0.08	0.018	0.04

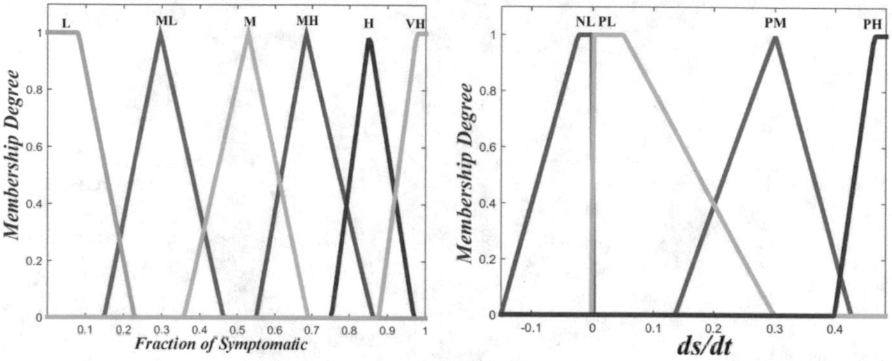

Fig. 3.15 Membership functions of the input and output variables

Peterman's data [39] has been adjusted through a best-fit procedure to find the value for the parameter m, resulting in $m = 0.237$ per year. Therefore, obtaining the solution of Anderson's model.

3.3.3   Type-1 p-Fuzzy System Evolution of Symptomatic HIV-Seropositive Population

The type-1 p-fuzzy model has been built from a FRBS encountered in [48], where the Mamdani inference method with the defuzzification of center of gravity is applied. Influenced by this information extracted from the literature, and using **Decision 2** in the decision block of Fig. 2.9, has been constructed as a type-1 p-fuzzy system , based on the direction field of the associated ordinary differential equation (3.6). The input variable is the fraction of the symptomatic population, s, which linguistic terms are: Low (L), Medium Low(ML), Medium (M), Medium High (MH), High(H), and Very High(VH). The output variable is the variation of the fraction symptomatic population $\dfrac{ds}{dt}$ with linguistic terms: Negative Low (NL), Positive Low (PL), Positive Medium (PM), Positive High (PH). The membership functions of the input and output variables are shown in Fig. 3.15.

The fuzzy rule base is in the following:

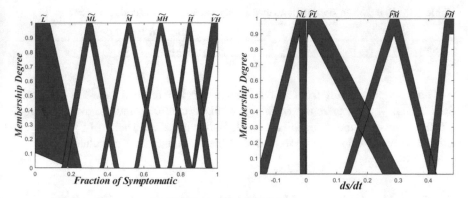

Fig. 3.16 FOU of the input and output variables

$$
\begin{aligned}
&\text{If } s \text{ is L} \quad \text{then} \quad \frac{ds}{dt} \text{ is PL;}\\
&\text{If } s \text{ is ML} \quad \text{then} \quad \frac{ds}{dt} \text{ is PM;}\\
&\text{If } s \text{ is M} \quad \text{then} \quad \frac{ds}{dt} \text{ is PH;}\\
&\text{If } s \text{ is MH} \quad \text{then} \quad \frac{ds}{dt} \text{ is PM;}\\
&\text{If } s \text{ is H} \quad \text{then} \quad \frac{ds}{dt} \text{ is PL;}\\
&\text{If } s \text{ is VH} \quad \text{then} \quad \frac{ds}{dt} \text{ is NL.}
\end{aligned}
\tag{3.7}
$$

Numerical simulations are carried out from an initial value of the fraction of symptomatic population $s(0) = 0$. Continuing with the flowchart of Fig. 2.9, the numerical integration used is the Trapezoid Rule [9]. The fraction of asymptomatic population, $a(t)$, is calculated using the equation $a(t) = 1 - s(t)$.

3.3.4   Interval Type-2 p-Fuzzy System Evolution of Symptomatic HIV-Seropositive Population

Using the information obtained from the aforementioned type-1 p-fuzzy system, an interval type-2 p-fuzzy system is constructed. The interval type-2 fuzzy rule base is the same as used in the type-1 fuzzy rule base, shown in (3.7), with the substitution of the name of the type-1 fuzzy set, that represents the linguistic term, for the corresponding name of the interval type-2 fuzzy set. In fact, notice that in Fig. 3.16 the FOUs of input and output variables have the name with the notation interval type-2 fuzzy set. These FOUs have been obtained after several empirical tests, selecting the ones that best approached the current data.

Table 3.13 Comparison in the performance among different models through the error

	Type-1 p-fuzzy	Interval Type-2 p-fuzzy
Error	0.1397	0.1055

Three outputs result from the inference for the interval type-2 FRBS (see Sect. 2.2.1): y_L, y_R, extremes of the generalized centroid, and the average value, y', that represents the defuzzified output. Following the flowchart of Fig. 2.9 and using Trapezoid Rule [9], the next value of the three outputs is calculated.

3.3.5   p-Fuzzy Systems Versus Data Peterman: A Comparison

A comparison of the type-1 and interval type-2 p-fuzzy systems performance with respect to the data of Peterman, numerical integrated, is developed through the maximum absolute error [9] at each data instant, calculating

$$error = \max_{i=1,\ldots,11} (|TIM(t_i) - DP(t_i)|),$$

where TIM is a p-fuzzy model of type-I, $(I = 1, 2)$, DP is the data of Peterman, and t_i is the instant (years after infection). The results are shown in Table 3.13. Notice that the accuracy of the interval type-2 p-fuzzy system to Peterman's data, numerical integrated, is superior to the type-1 p-fuzzy system.

Depicted in Fig. 3.17 are outputs generated by the type-1 and interval type-2 p-fuzzy systems, for the values y', y_L, and y_R, namely $s(t)$, $s_L(t)$, and $s_R(t)$, the latest being the curves that compose a range that contains the Peterman's data, after a numerical integration, and the type-1 p-fuzzy system. Notice that the numerical integration value $s_R(t)$ is limited by 1 from $t_0 = 3.72$ because the fraction of symptomatic population belongs to interval [0, 1].

3.3.6   Conclusion

In this section, p-fuzzy systems have been used as an alternative methodology to understand phenomena that present uncertainty. Using information from the literature and taking **Decision 2** of Fig. 2.9 flowchart, a type-1 p-fuzzy system to model the transference of HIV-seropositive population fraction from asymptomatic to symptomatic has been built. Motivated by this method, an interval type-2 p-fuzzy system is developed, obtaining a range determined for the numerical integration of the defuzzified outputs, smallest and biggest centroids of type-1 fuzzy sets embedded in the interval type-2 fuzzy sets, from the Mamdani inference method. In this range are included the type-1 p-fuzzy system approximation and the Peterman's data. A comparison is made among the two approximations obtained through both p-fuzzy systems, type-1 and interval type-2, using the maximum absolute error,

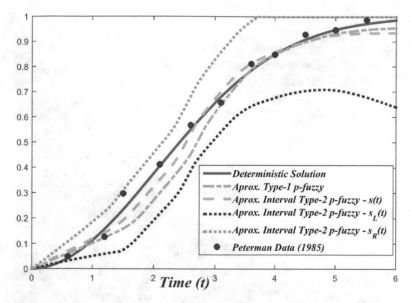

Fig. 3.17 Symptomatic HIV-seropositive population interpreted by the different models

showing that the interval type-2 fuzzy method is more accurate than the type-1 fuzzy technique.

3.4   Cellular Automata and Interval Type-2 Fuzzy Approach: A Partnership in a Study of HIV-Seropositive Population

The main idea of Cellular Automata (CA) models is to consider each position (or region) of a spatial domain as a cell in which is attributed a certain state. The state of each cell is modified according to its own state and the states of its neighbor cells. These states are correlated through a number of simple rules that imitates the biological and physical laws [15]. Presented in this section are two different CA models with inputs coming from two interval type-2 FRBS, using the Mamdani inference method and the centroid as the defuzzification technique. The first CA has been built from an adaptation of the system of ordinary differential equations proposed by Anderson et al. [3], by adding three phases of viral evolution of the HIV infection in the absence of antiretroviral treatment. In this CA individuals represented by cells coexist artificially: susceptible, individuals who are in the period of acute infection, asymptomatic, and symptomatic individuals. Two interval type-2 FRBS control the transformation of a cell as a function of time and neighboring cells. These interval type-2 FRBS have inputs depending on time and outputs on the infection risk at the acute and asymptomatic phase. The output model favorably compares to the data from 1980 to 1995, when the use of

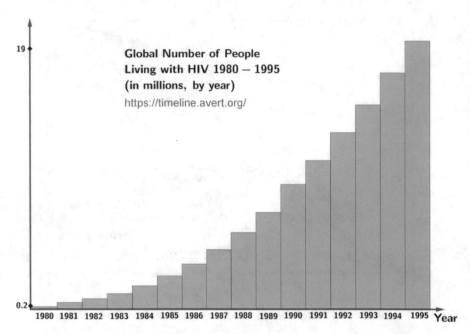

Fig. 3.18 Evolution of the number of individuals infected with HIV-seropositive in the period 1980 through 1995, in millions by year, in the world. Data extracted from [5] to build the bar chart in GeoGebra [20]

antiretroviral drugs was not available [26]. The second CA model has been built from a system of ordinary differential equations to describe the dynamics of transfer and return of HIV-seropositive population, from asymptomatic to symptomatic and vice versa, when receiving antiretroviral treatment. In this CA model, asymptomatic and symptomatic individuals coexist with their respective rates of transference and return, which are the outputs of type-1 FRBS or of an interval type-2 FRBS. Rates depend on adherence to treatment, viral load, and $CD4+$ T lymphocyte, recalling that this is the main cell attacked by the HIV. Both CA models are shaped like a torus, being two-dimensional with Moore neighborhood [34] and periodic (toroidal) boundary [42].

3.4.1   Cellular Automaton for Epidemiological Modeling of HIV Infection

An epidemiological model using a CA is developed, based on the work of Jafelice et al. [26]. The graph of the evolution of the number of individuals contaminated with HIV in the world from 1980 until 1995, when the antiretroviral treatment was not applied, is shown in Fig. 3.18.

The original features of the proposed methodology in [26] are three fold:

- Incorporates to the Anderson model [3] recent understanding of the three phases of HIV infection progression under the condition of no antiretroviral drug use, see Fig. 3.19.
- The use of a CA to model the HIV progression in a population without treatment.
- The use of an interval type-2 FRBS to control the transformation of a cell according to time and neighboring cells. This simulates the progression in time and transmission via contact among the infected population.

The modeling of the epidemiological system described by the following system of differential equations, in Eq. (3.8), takes into account the stages of the HIV infection described below [1]:

1. Acute HIV Infection: Acute HIV infection is the earliest stage of HIV infection. During this time, some individuals have flu-like symptoms, such as fever, headache, and rash. In the acute stage of infection, HIV multiplies rapidly and spreads throughout the body. The level of HIV in the blood is very high, which increases the risk of HIV transmission.
2. Asymptomatic Phase: In this second stage of HIV infection the HIV continues to multiply in the body but at very low levels. Individuals with chronic HIV infection may not have any HIV-related symptoms.
3. Symptomatic Phase: AIDS is the final, most severe stage of HIV infection. Because HIV has severely damaged the immune system, the body cannot fight off opportunistic infections. Once a person is diagnosed with AIDS, they can have a high viral load and are able to transmit HIV to others very easily.

The model that considers the stages of HIV infection is given by

$$\frac{dh}{dt} = B - \mu h - \lambda_1 rh - \lambda_2 rh \qquad \lambda_1 = \beta_1 \frac{i}{h + i + a}$$

$$\frac{di}{dt} = \lambda_1 rh + \lambda_2 rh - (q + \mu)i \qquad \lambda_2 = \beta_2 \frac{a}{h + i + a}$$

$$\frac{da}{dt} = (1 - p)qi - (\mu + k)a$$

$$\frac{ds}{dt} = pqi + ka - (d + \mu)s \qquad (3.8)$$

$$n(t) = h(t) + i(t) + a(t) + s(t),$$

where the variables are numbers of susceptible individuals, $h(t)$, individuals in the acute infection phase, $i(t)$, asymptomatic individuals, $a(t)$, AIDS patients, $s(t)$, and $n(t)$ is the total number of individuals.

The parameters are:

- B: immigration rate of susceptible individuals per year;
- μ: yearly population mortality rate due to natural causes;

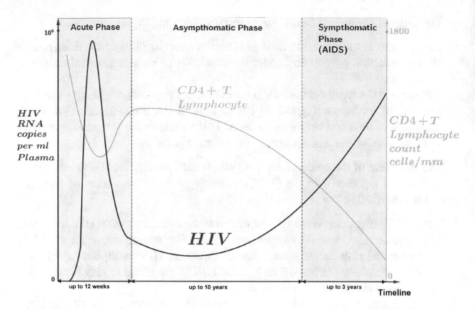

Fig. 3.19 Schematic view of the currently accepted natural history of HIV infection [14, 38] and [43]

- λ_1: probability of acquiring the infection through acute-infected sexual partners chosen at random;
- β_1: probability of acquiring infection from a partner with acute infection;
- λ_2: probability of acquiring infection through random asymptomatic sexual partners;
- β_2: probability of acquiring infection from an asymptomatic partner;
- r: rate that represents the number of sexual partners per year;
- q: percentage of individuals that go from the acute to the asymptomatic phase per year;
- p: proportion of individuals that go from the acute infection phase to develop AIDS;
- k: percentage of individuals per year who move from asymptomatic to symptomatic phase;
- d: percentage of the symptomatic population that dies due to AIDS per year.

The CA's rules used in this model are established taking in account the characteristics of system Eq. (3.8) such that:

1. The simulation of the cellular automaton is performed on a grid 60×60 cells with a maximum of 40 iterations possible. The initial number of susceptible individuals is 713, one individual in the acute phase of infection, 36 asymptomatic individuals, and none symptomatic individual, which are all placed randomly in the computational environment.

Fig. 3.20 The alternatives of movements for the individuals inside the cellular automaton [44]

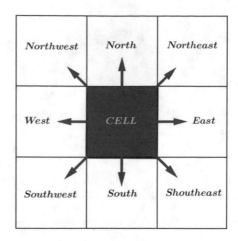

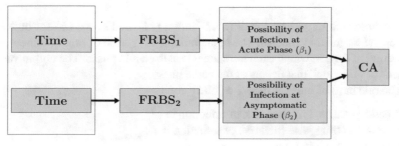

Fig. 3.21 Schematic representation of the methodology utilized to obtain the CA parameters

2. Each cell has eight degrees of freedom, i.e., it can move towards north, south, east, west, northeast, northwest, southeast, and southwest, as shown in Fig. 3.20.
3. Any susceptible individual, after a certain number of iterations, randomly chooses an individual in order to simulate a sexual intercourse. From then on, there is a risk of infection if this individual is at the acute or at the asymptomatic phase of infection. The risk of infection is determined through two interval type-2 FRBS, called $FRBS_1$ and $FRBS_2$. Their outputs are β_1 and β_2, respectively, which are the inputs of the CA, as shown in Fig. 3.21.
4. If the risk of an infection through a sexual intercourse is less than or equal to a randomly chosen number between 0 and 1, then the susceptible individual is infected.
5. When a cell representing an individual at the acute phase of infection reaches the period of time sufficient for the change of stage, it may become asymptomatic or pass directly to the symptomatic phase, using a predefined probability.

Both interval type-2 FRBS used for the CA feed have their membership functions' support based on biological phenomena, the reasons that are detailed in [26]. Shown in Fig. 3.22 are the FOU of the linguistic terms associated with the interval type-2 $FRBS_1$, referred as the input and output variables.

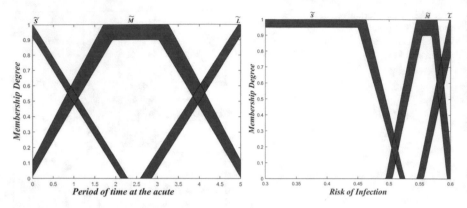

Fig. 3.22 FOU of the input and output variables

The linguistic terms are: Small ($\widetilde{S}$), Medium ($\widetilde{M}$), and Large ($\widetilde{L}$) for both input and output linguistic variables. The input variable is "period of time" that represents the period of time in which each individual is at the acute phase. The output variable is "risk of infection" that occurs at the acute phase.

The rule base of interval type-2 $FRBS_1$ are:

- If period of time is $\widetilde{S}$ then risk of infection is $\widetilde{M}$;
- If period of time is $\widetilde{M}$ then risk of infection is $\widetilde{L}$;
- If period of time is $\widetilde{L}$ then risk of infection is $\widetilde{M}$.

Illustrated in Fig. 3.23 is the risk of infection as a function of the period of time obtained from the interval type-2 $FRBS_1$. When there is a possibility of a symbolic sexual intercourse between a susceptible individual and an individual in the period of acute infection, the CA checks the acute phase of infection's period of time through the interval type-2 $FRBS_1$, obtained all the values of input's support, as depicted in Fig. 3.23.

Using the dynamics of item 5 of the CA's characteristics, a decision whether the susceptible individual is infected or not is made by the CA. Note that the graph in Fig. 3.23 is similar to the qualitative behavior of the viral load at acute phase of the infection as that of the first weeks, depicted in Fig. 3.19.

The input variable's FOUs of the interval type-2 $FRBS_2$ are shown in Fig. 3.24, representing the period of time that an individual is at the asymptomatic phase. The output variable is the same as the interval type-2 $FRBS_1$ described in Fig. 3.22.

The rule base of interval type-2 $FRBS_2$ is given by:

- If period of time is $\widetilde{S}$, then risk of infection is $\widetilde{S}$;
- If period of time is $\widetilde{M}$, then risk of infection is $\widetilde{S}$;
- If period of time is $\widetilde{L}$, then risk of infection is $\widetilde{M}$.

Using the scheme of Fig. 3.21, the same procedure for the infection of susceptible individuals by individuals at the acute phase is performed with the values obtained

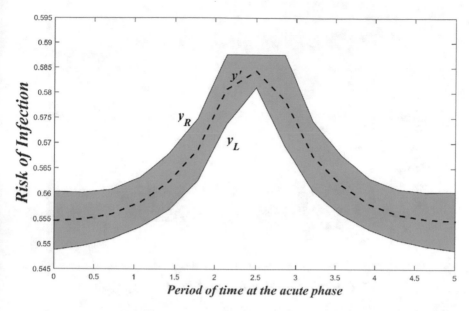

Fig. 3.23 Graph of the risk of infection versus the period of time at the acute phase of an individual

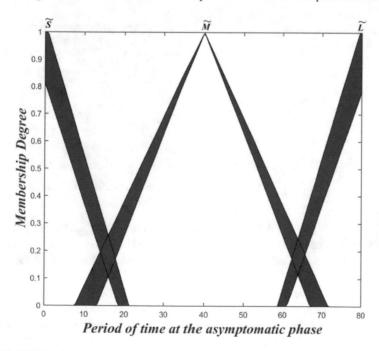

Fig. 3.24 FOU for the period of time

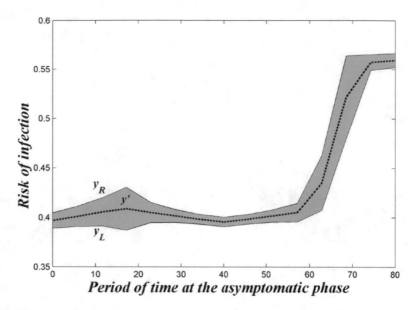

Fig. 3.25 Graph of risk of infection versus the period of time at the asymptomatic phase of an individual

Fig. 3.26 A snapshot of the CA model output

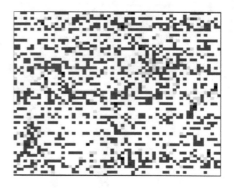

from the interval type-2 $FRBS_2$ for asymptomatic individuals, shown by the graph of Fig. 3.25. Notice that the behavior of the viral load in an asymptomatic individual shown in Fig. 3.25 is similar to the qualitative behavior of the graph of Fig. 3.19 at the same phase.

A snapshot of the CA model is shown in Fig. 3.26: the brown background denotes the computational environment in which individuals live; the white squares are the susceptible individuals; the black squares are the symptomatic individuals; the individuals at the acute phase are represented by the red squares and the asymptomatic individuals by the yellow squares.

The CA outputs provide the evolution of the number of individuals as function of time in the four classes given by: susceptible, at the acute phase, asymptomatic, and symptomatic. The output of each class is acquired using the three values, y_L,

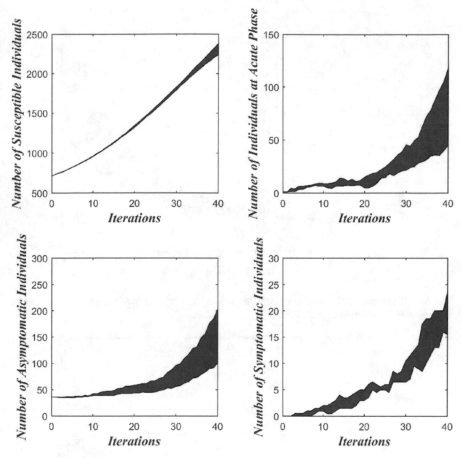

Fig. 3.27 Evolution on time of the four classes of populations (CA outputs) considering the maximum and minimum of the numbers of the individuals

y_R, and y', obtained through the interval type-2 $FRBS_1$ and $FRBS_2$. The graph of Fig. 3.27 is composed of the four outputs of the CA, corresponding to the classes of individuals at each iteration, where the maximum and the minimum of the numbers of individuals is calculated. Notice that a range of values is obtained for the CA output.

The total number of HIV carriers is the sum of the asymptomatic, symptomatic, and the individuals at the acute phase. Shown in Fig. 3.28 is the graph of the HIV carriers number acquired from the CA's outputs as previously described, and a range of values determined by these numbers.

Consistent with the graph of Fig. 3.18 is the bar chart of Fig. 3.29, which indicates that the CA model tracks the historical data, which validated the mathematical modeling made through this study.

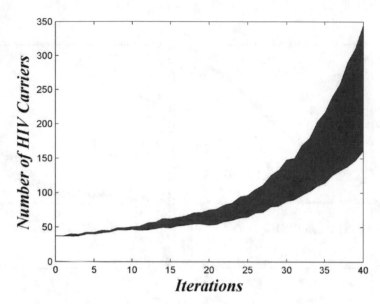

Fig. 3.28 Evolution of the number of HIV carriers in CA output, considering the maximum and minimum of the numbers of individuals

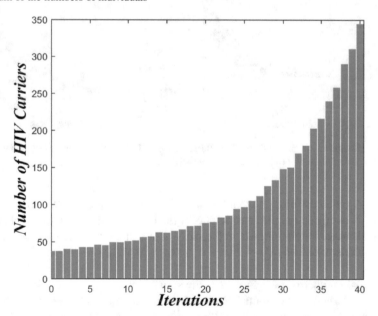

Fig. 3.29 Bar graph of the evolution of the number of HIV carriers in CA output, considering the maximum of the numbers of individuals

3.4.2 Cellular Automaton of the HIV-Seropositive Population Dynamics with Antiretroviral Therapy

Inspired by the works [3, 23] and [37], the HIV-seropositive evolution model with antiretroviral treatment are presented, described by the equations:

$$
\begin{aligned}
\frac{da}{dt} &= -\lambda(v, c, f)\, a + \gamma(v, c, f)\, s - \mu a \qquad a(0) = a_0 \quad 0 < a_0 < 1 \\
\frac{ds}{dt} &= \lambda(v, c, f)\, a - \gamma(v, c, f)\, s - (\mu + d)\, s \quad s(0) = s_0 = 1 - a_0 \\
a + s &= 1,
\end{aligned}
$$

$$(3.9)$$

where the variables and parameters involved are:

- a: fraction of asymptomatic individuals as a function of time;
- s: fraction of symptomatic individuals as a function of time;
- λ: transference rate of the fraction of the population from asymptomatic to symptomatic depending on the viral load, v, the $CD4+$ T lymphocyte level, c, and the adherence to treatment, that is, follow the medication administration protocol, f;
- γ: rate of return from the fraction of the symptomatic to asymptomatic population, which depends on the v, c and f;
- μ: percentage of mortality from natural causes of the two fractions of individuals;
- d: percentage of mortality of the fraction of symptomatic individuals due to the AIDS.

A type-1 FRBS has been built to determine λ and γ, which are the type-1 FRBS outputs, using the Mamdani inference method and the center of gravity defuzzification technique. The transference and return of the fraction of the HIV-seropositive population have been modeled according to the viral load, $CD4+$ T lymphocyte level, and adherence to treatment, according to the suggestion of specialists in the field. The membership function's supports of the linguistic terms of the input variables, v, c, and f have been obtained through the literature information [23], being standardized with the intent of facilitating the modeling. The same technique is applied for the construction of the membership functions of the output variables.

The input variables have the following linguistic terms:

- v: Low (L), Medium (M) and High (H);
- c: Very Low (VL), Low (L), Medium (M), Medium High (MH), and High (H);
- f: Non-Adequate (NA) and Adequate (A).

Output variables are the same linguistic terms: Weak (W), Medium (M), and Strong (S). The membership functions of the input variables v, c, and f, and of outputs λ and γ are shown in Fig. 3.30. Presented in Tables 3.14 and 3.15 are the rule base of the type-1 FRBS.

Table 3.14 Rule base for non-adequate adherence (f)

v	L	M	H	v	L	M	H
c				c			
VL	S	S	S	VL	W	W	W
L	M	S	S	L	W	W	W
M	M	M	M	M	W	W	W
MH	MW	MW	M	MH	MW	MW	W
H	W	W	M	H	M	M	W
(a) Transference rate (λ)				(b) Return rate (γ)			

In order to solve the system (3.9), the values $v = 0.5$, $c = 0.46$, and $f = 0.5$ have been chosen, for which $\lambda = 0.298$ and $\gamma = 0.236$ are obtained through the type-1 FRBS previously built. The parameters μ, d and, the initial conditions, a_0, b_0 have been extracted from the literature [37]. Thus, the system (3.9) is numerically solved with the parameters and the initial conditions detailed in Table 3.16 using the 4th Order Runge-Kutta method [4].

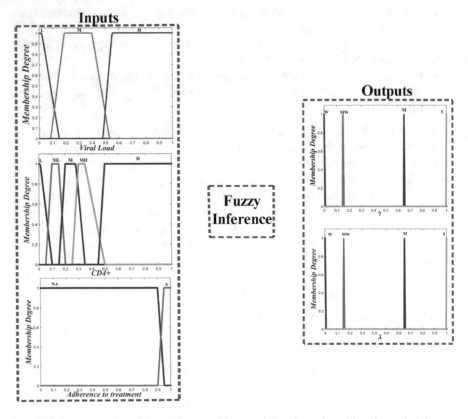

Fig. 3.30 The type-1 FRBS for HIV-seropositive population dynamics with antiretroviral therapy

Table 3.15 Rule base for adequate adherence (f)

v	L	M	H	v	L	M	H
v				c			
VL	M	M	M	VL	W	W	W
L	MW	M	M	L	MW	W	W
M	MW	MW	MW	M	MW	MW	MW
MH	W	W	MW	MH	M	M	MW
H	W	W	MW	H	S	S	MW
(a) Transference rate (λ)				(b) Return rate (γ)			

Table 3.16 Parameters and initial conditions used in the numerical resolution of the system (3.9)

Parameter	Value
a_0	0.8
s_0	0.2
λ	0.298
γ	0.236
μ	0.031
d	0.05

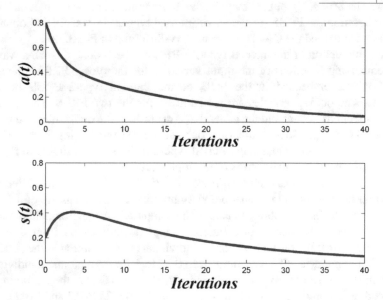

Fig. 3.31 Numerical solution of the system (3.9)

Described in Fig. 3.31 are the graphs of the approximations of the system solution (3.9).

Motivated by the system (3.9) a CA has been built with rules compatible with the system's dynamics. In the CA, asymptomatic and symptomatic HIV-seropositive individuals interact artificially. The CA starts with asymptomatic individuals (80 % of the total) and the rest are symptomatic presenting each of these 8 degrees of freedom, as shown in Fig. 3.20. When asymptomatic, they are liable to become

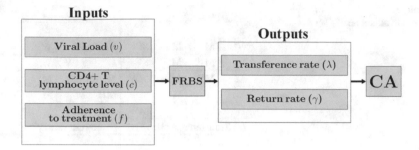

Fig. 3.32 Schematic representation of the methodology utilized to obtain λ and γ, the CA's inputs

symptomatic at a rate proportional to $\lambda(v, c, f)$, obtained by the FRBS (of type-1 or interval type-2), or to die due to natural causes. When symptomatic, they can return to the asymptomatic phase at a rate proportional to $\gamma(v, c, f)$, obtained in the same way as $\lambda(v, c, f)$, or they may die due to immunodeficiency or natural causes. In each iteration the FRBS (of type-1 or interval type-2) is consulted to obtain the rates λ and γ to feed the CA. This dynamic is illustrated in Fig. 3.32.

The construction of the interval type-2 FRBS has been done in such a way that the membership functions of the input variables for the type-1 FRBS, previously described, are embedded in the FOU of the interval type-2 FRBS linguistic variables. The output variables are the same as for the two FRBS. The inference system is from Mamdani and the generalized centroid is the defuzzification method (Sect. 2.2.1). As previously, the rule base is built from Tables 3.14 and 3.15 substituting for the corresponding interval type-2 fuzzy sets. Illustrated in Fig. 3.33 are the FOU of the linguistic terms of the input variables.

The CA simulation has been performed on a 25×25 cells grid with 40 iterations. The initial number of asymptomatic and symptomatic individuals are 368 and 92, respectively. The individuals are placed in the computational environment randomly, as well as their age that is distributed among the individuals from 0 to 100. Adherence to treatment is random for all individuals with variation in the [0.5, 0.8], $CD4+$ T lymphocyte level is random from 0.35 to 1 for asymptomatic individuals, and from 0.1 to 0.35 for symptomatic individuals. Likewise, at the beginning, the viral load of asymptomatic individuals is random from 0.25 to 0.5 and from 0.8 to 1 for symptomatic individuals. The natural death of asymptomatic and symptomatic individuals is performed based on the iteration that they reach an age greater than 100. For each symptomatic individual, in each iteration, a random value is chosen in the interval [0, 1], and when this value is less than or equal to 0.03, the individual dies from immunodeficiency.

A snapshot of the CA is shown in Fig. 3.34, the green background denotes the computational environment in which individuals live; the white squares are the asymptomatic individuals; the red squares are the symptomatic individuals.

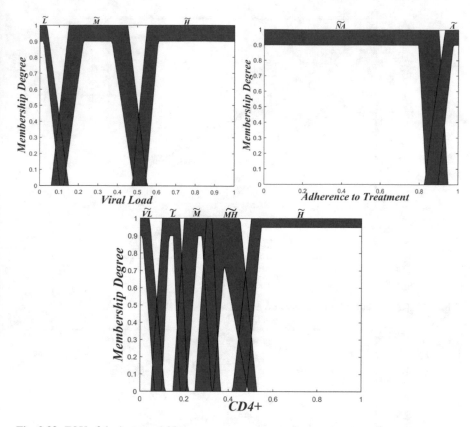

Fig. 3.33 FOU of the input variables

Fig. 3.34 A snapshot of the
CA model output

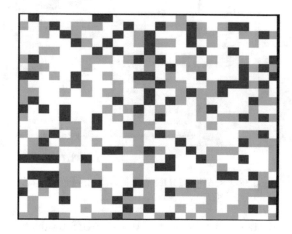

The graph of Fig. 3.35 represents the CA outputs of the asymptomatic fraction
using the values of the rates $\lambda(v, c, f)$ and $\gamma(v, c, f)$ obtained from the type-1 and
interval type-2 FRBS, namely $\widehat{a}$, $\widehat{a}_L$, and $\widehat{a}_R$. Shown in Fig. 3.36 are the curves

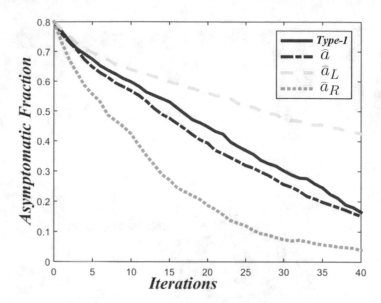

Fig. 3.35 CA output for the fraction of asymptomatic individuals

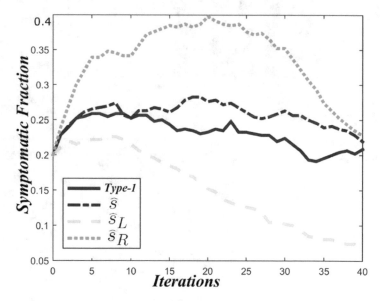

Fig. 3.36 CA output for the fraction of symptomatic individuals

resulting from the interval type-2 FRBS for the fraction of symptomatic individuals using the same dynamics, namely $\widehat{s}, \widehat{s}_L$, and $\widehat{s}_R$.

Note that the graphs of Figs. 3.35 and 3.36 are consistent with the numerical solution shown in Fig. 3.31.

3.4.3 Conclusion

The partnership of cellular automaton with the type-1 FRBS and interval type-2 FRBS allows each cell to represent an individual carrying the virus endowed with proper characteristics. An important fact to be noted is that a cellular automaton with simple rules, as those used in the simulations of the study, can capture a more complex behavior which is closer to the reality. Shown in Sect. 3.4.1 is an artificially simulated system in which individuals behave consistently with the evolution of the number of HIV carriers in the world, is shown in Fig. 3.18.

In fact, the combinations of the mathematical procedures provided a quality approach to epidemic modeling, that favorably compares to the statistical data from the period 1980 to 1995, when antiretroviral treatment was not used. Furthermore, as a result of the cellular automaton's randomness and the features of the interval type-2 FRBS, is the attainment of an infection range of possibilities for the acute and asymptomatic phases; compatible with the most accepted natural history of HIV infection, as shown in Fig. 3.19. These two stand out features are points of validation of the methodology proposed.

The cellular automaton presented in Sect. 3.4.2 models a fictitious transfer and return of asymptomatic to symptomatic HIV-seropositive individuals via parameters of type-1 and interval type-2 FRBS, with antiretroviral treatment. Note that, as first characteristic to be highlighted, the CA output with the parameters obtained by the type-1 FRBS are contained in the CA outputs in the range generated through the y_L and y_R, outputs from the interval type-2 FRBS. As a consequence of the interval type-2 fuzzy approach that has been computationally implemented, the membership functions of the input variables of the type-1 FRBS are embedded in the FOU of the input variables of the interval type-2 FRBS. Therefore, the outputs of the CA obtained from type-1 and interval type-2 FRBS preserve the same characteristic as shown in Figs. 3.35 and 3.36. In addition, as a second characteristic, the similar qualitative behavior of these graphs stands out to those of the numerical solution of the system shown in Fig. 3.31. Both highlighted properties provided a quality approach to the epidemic modeling.

3.5 Population Models Through Interval Type-2 p-Fuzzy Systems

Initially in this section, a p-fuzzy model of Malthus [25] is obtained through a type-1 and an interval type-2 FRBS using the Mamdani inference method. The defuzzification method for the type-1 approach is the centroid and for the interval type-2 is used the interval type-2 fuzzy logic toolbox of Castro et al. [11] to obtain the generalized centroid. Both results are compared to best fit with the Malthus's deterministic curve [32]. Another alternative for the Malthusian mathematical modeling is using an ANFIS training (see Sect. 2.1.2) to obtain a type-

1 p-fuzzy system [17], which is the influence for the construction of an interval type-2 p-fuzzy system in order to check which one best matches the Malthus's deterministic solution. Afterwards, a data set that contains information about Peru's population from 1961 to 2016 is utilized for the next population model [25]. Then, acquired from an ANFIS training, a type-1 p-fuzzy system is built, followed by the development of an interval type-2 p-fuzzy system. This system is based in its type-1 counterpart and it is submitted for a comparison with the specific Peru's population data. In all these studies a maximum relative error, calculated in a corresponding instant according with the model, is applied for comparison.

3.5.1　Interval Type-2 p-Fuzzy System for the Malthus Model: Mamdani Inference Method

The first mathematical modeling in this set of population studies is the Malthus' deterministic model [32] given by

$$\begin{cases} \dfrac{dP}{dt} = \eta P \\ P(0) = P_0, \end{cases} \tag{3.10}$$

where P is the population at instant t and η is the constant rate of population growth. The solution of Eq. (3.10) is given by $P(t) = P_0 \exp(\eta t)$. Considering $\eta = 0.47$ and $P_0 = 2$ the deterministic solution is given by $P(t) = 2\exp(0.47t)$.

In [7] a type-1 p-fuzzy system for the Malthus model is presented, with the input variable P and the output variable $\dfrac{dP}{dt}$, using the Mamdani inference method. Based on this information, applying the **Decision 2** of the p-fuzzy flowchart of Fig. 2.9 decision block, a type-1 FRBS is constructed with the linguistic terms of the two variables defined by: Very Low (VL), Low (L), Medium (M), and High (H). The membership functions of the input and output variables are shown in Fig. 3.37.

The rule base is elaborated based on the behavior of the direction field Eq. (3.10) given by:

- If P is VL, then $\dfrac{dP}{dt}$ is VL;
- If P is L, then $\dfrac{dP}{dt}$ is L;
- If P is M, then $\dfrac{dP}{dt}$ is M;
- If P is H, then $\dfrac{dP}{dt}$ is H.

Following the flowchart of the p-system in Fig. 2.9, the Trapezoid Rule [9] is used as a numerical integration method to obtain the next population value.

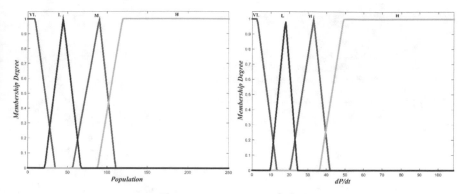

Fig. 3.37 Linguistic variables of the type-1 FRBS

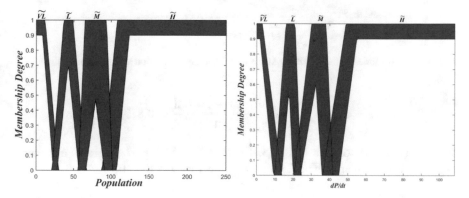

Fig. 3.38 Linguistic variables of the interval type-2 FRBS

Based on this system an interval type-2 FRBS has been built. The graphs of the FOU of the input variable P and output, $\dfrac{dP}{dt}$, are shown in Fig. 3.38. As aforementioned, linguistic terms and the rule base of the type-1 FRBS are replaced for the corresponding interval type-2 fuzzy sets.

The interval type-2 FRBS has been constructed using the Mamdani inference method (see Sect. 2.1.1), following the flowchart of Fig. 2.9 and using the Trapezoidal Rule [9] as a method of numerical integration to find the corresponding solution.

Shown in Fig. 3.39 are graphs as a function of time for each mathematical modeling's results: the type-1 p-fuzzy system, the deterministic, and the interval type-2 p-fuzzy system, respectively.

Next, a comparison is done among the trajectories of the two types of p-system and the IVP solution through the accuracy measure $error_1$ and $error_2$ given by

$$error_1 = \max_{t \in [0,10]} \frac{|P(t) - P_{f_1}(t)|}{|P(t)|} = 1.038,$$

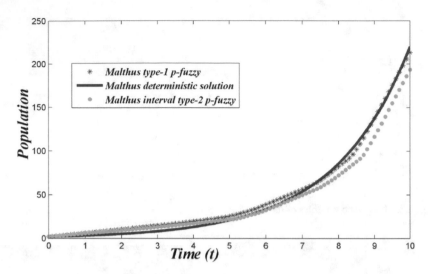

Fig. 3.39 Trajectories of the p-fuzzy systems and the analytic solution as a function of time

$$error_2 = \max_{t \in [0,10]} \frac{|P(t) - P_{f_2}(t)|}{|P(t)|} = 0.819,$$

where $P(t)$ is the solution of Eq. (3.10), $P_{f_1}(t)$ is the type-1 p-fuzzy's trajectory, and $P_{f_2}(t)$ is the interval type-2 p-fuzzy's trajectory, $t \in [0, 10]$.

3.5.2   Interval Type-2 p-Fuzzy System for the Malthus Model: Takagi–Sugeno–Kang Inference Method

In this section, a mathematical modeling of the p-fuzzy system based on the Malthusian principle [32] for population growth:

> "the variation of a population is proportional to the population at each instant." This principle determines that the population variation grows as the population grows.

Based on this information from the literature and **Decision 1**, consider the four following features for Malthus p-fuzzy model:

- the constant of proportionality, 0.47;
- the initial population of two individuals;
- total population when it reaches instant 10 is 250;
- a type-1 FRBS whose output is the population variation rate.

An ANFIS training is done upon the interval $D = [2, 250]$ as the input variable domain and the output variable defined in $0.47D$. The system is constructed with the linguistic input variable, population P, and the output, $\dfrac{dP}{dt}$. The linguistic terms for the population are: Low (L), Medium Low (ML), Medium High (MH), and High (H), represented in Fig. 3.40 by triangular membership functions.

The output variable is a polynomial of form $f(P) = mP + n$, where the parameters m and n are determined by ANFIS. The rule base is composed by:

- If P is L, then $\dfrac{dP}{dt} = 0.47P + 3.85 \cdot 10^{-5}$;
- If P is ML, then $\dfrac{dP}{dt} = 0.47P + 0.00163$;
- If P is MH, then $\dfrac{dP}{dt} = 0.47P + 0.003221$;
- If P is H, then $\dfrac{dP}{dt} = 0.47P + 0.04812$.

Following the flowchart of Fig. 2.9, the first output from the initial population of 2 individuals is calculated. Next, the numerical integration method of Simpson's Rule [9] is applied to obtain the next value $P_{f_3}(t)$ of the population corresponding to this type-1 p-fuzzy trajectory, up to $t = 10$.

Inspired by the type-1 FRBS, an interval type-2 FRBS has been constructed which input variable is depicted in Fig. 3.41. Note that the type-1 linguistic terms of the input variable are substituted for the counterpart interval type-2 fuzzy set.

The rule base is given by:

- If P is $\widetilde{L}$, then $\dfrac{dP}{dt} = [0.46,\ 0.48]P + [-0.0099,\ 0.01]$;

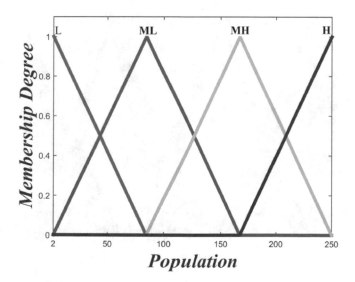

Fig. 3.40 Membership functions for the input variable P

- If P is $\widetilde{ML}$, then $\dfrac{dP}{dt} = [0.44,\ 0.5]P + [-0.0283,\ 0.0316]$;
- If P is $\widetilde{MH}$, then $\dfrac{dP}{dt} = [0.45,\ 0.49]P + [-0.0167,\ 0.0232]$;
- If P is $\widetilde{H}$, then $\dfrac{dP}{dt} = [0.44,\ 0.5]P + [-0.2951,\ 0.30482]$.

An interval type-2 FRBS is established by the TSK inference method (see Sect. 2.1.1), using the flowchart of Fig. 2.9, the defuzzified output given by Eq. (2.20). Applying the Simpson Rule [9] as a method of numerical integration the next value of the population is obtained.

Using the deterministic solution of the Malthus given by $P(t) = 2\exp(0.47t)$, are calculated $error_3$ and $error_4$ defined by

$$error_3 = \max_{t \in [0,10]} \frac{|P(t) - P_{f_3}(t)|}{|P(t)|} = 0.1228,$$

$$error_4 = \max_{t \in [0,10]} \frac{|P(t) - P_{f_4}(t)|}{|P(t)|} = 0.1226,$$

where $P(t)$ is the solution of Eq. (3.10), $P_{f_3}(t)$ is the type-1 p-fuzzy trajectory, and $P_{f_4}(t)$ is the interval type-2 p-fuzzy trajectory, $t \in [0,\ 10]$.

Described in Fig. 3.42 are graphs of the mathematical modeling's results as a function of time corresponding to: the type-1 p-fuzzy system, the deterministic, and the interval type-2 p-fuzzy system, respectively.

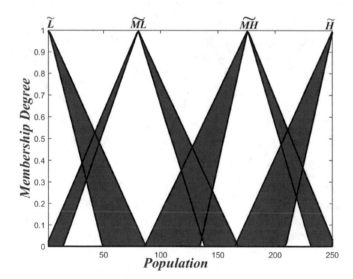

Fig. 3.41 FOU of the input variable

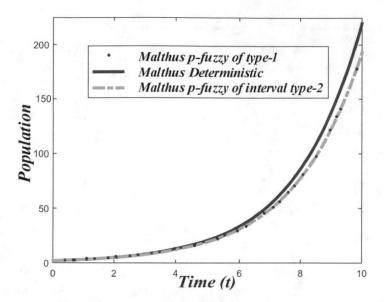

Fig. 3.42 Trajectories of the p-fuzzy systems and the analytic solution as a function of time

Table 3.17 Peru's data population $p(t_i)$

Year (t_i)	Population ($p(t_i)$)	$\dfrac{\Delta p}{\Delta t_i}$
1961	10,420,357	0.3364
1972	14,121,564	0.4045
1981	17,762,231	0.4064
1993	22,639,443	0.3986
2007	28,220,764	0.3757
2013	30,475,000	0.3378
2016	31,488,655	0.3113

3.5.3   Interval Type-2 p-Fuzzy System for Peruvian Population Data

The censuses of Peru's population from 1961 to 2016, shown in the two first columns of Table 3.17, have been used to build a type-1 FRBS, determined through ANFIS, using **Decision 1** from the decision block of the scheme in Fig. 2.9.

In the third column of Table 3.17 there is the approximation $\dfrac{\Delta p}{\Delta t_i} = \dfrac{p(t_{i+1}) - p(t_i)}{t_{i+1} - t_i}$, $i = 1, \ldots, 6$, and $\dfrac{\Delta p}{\Delta t_7} = \dfrac{p(t_7) - p(t_8)}{t_7 - t_8}$, $t_8 = 2017$, and $p(t_8)$ is the Peruvian population forecast for 2017 given by 31,800,000. Through an ANFIS training, feed by p and $\dfrac{\Delta p}{\Delta t_i}$, it is obtaining: the input membership functions, the

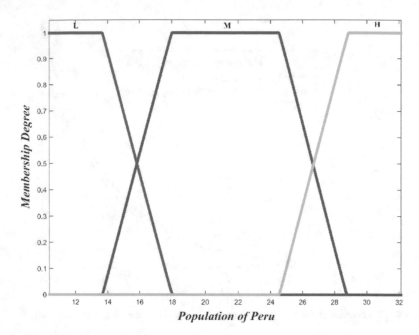

Fig. 3.43 Membership functions for the input variable p

output polynomials, and base rule. Notice that the range of the input variable is the interval [10.42, 32.11] in millions of population. For the population p, the linguistic terms are Low (L), Medium (M), and High (H), which membership functions are shown Fig. 3.43.

The rule base is composed by:

- If p is L, then $\dfrac{dp}{dt} = 0.0181p + 0.1478$;

- If p is M, then $\dfrac{dp}{dt} = -0.001515p + 0.433$;

- If p is H, then $\dfrac{dp}{dt} = -0.0192p + 0.9189$.

Furthermore, following the p-fuzzy flowchart of Fig. 2.9, a defuzzified output is acquired using the Simpson's Rule [9] numerical integration, to obtain the next value of the population.

Based on this information, an interval type-2 FRBS with TSK inference method (see Sect. 2.1.1) is built, which input variable is shown in Fig. 3.44. Note that the linguistic terms of the input type-1 FRBS have been substituted by the interval type-2 fuzzy set with the same label and a tilde on the top.

The rule base is giving by the following statements:

- If p is $\tilde{L}$, then $\dfrac{dp}{dt} = [0.0171, 0.0219]p + [0.1478, 0.1522]$;

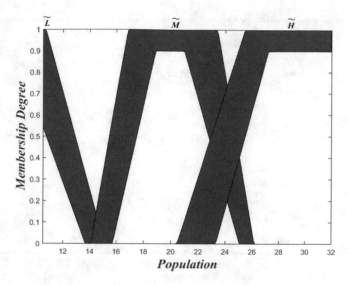

Fig. 3.44 FOUs of the input linguistic variable

- If p is $\widetilde{M}$, then $\dfrac{dp}{dt} = [-0.0115,\ -0.0015]p + [0.4331,\ 0.6069]$;
- If p is $\widetilde{H}$, then $\dfrac{dp}{dt} = [-0.0391,\ 0.0008]p + [0.9169,\ 1.0831]$.

To determine the trajectory of the interval type-2 p-fuzzy system has been utilized the numerical integration method of Simpson's Rule, applied to the output the interval type-2 FRBS. Shown in Fig. 3.45 is the population of Peru $p(t_i)$ of Table 3.17 and the graphs of the population as a function of time obtained through the type-1 and interval type-2 p-fuzzy systems. Through this study, it is possible to estimate the population values at each year, whereas the simulation is performed with $h = 1$ spacing.

Finally, comparing the data, $p(t_i)$, $i = 1, \ldots, 7$, with the trajectories obtained through both, type-1 and interval type-2 p-fuzzy systems, $error_5$ and $error_6$ are calculated:

$$error_5 = \max_{i=1,\ldots,7} \frac{|p(t_i) - p_{f_5}(t_i)|}{|p(t_i)|} = 0.0255, \tag{3.11}$$

$$error_6 = \max_{i=1,\ldots,7} \frac{|p(t_i) - p_{f_6}(t_i)|}{|p(t_i)|} = 0.0205, \tag{3.12}$$

where $p_{f_5}(t_i)$ and $p_{f_6}(t_i)$ are determined type-1 and interval type-2 p-fuzzy systems, respectively.

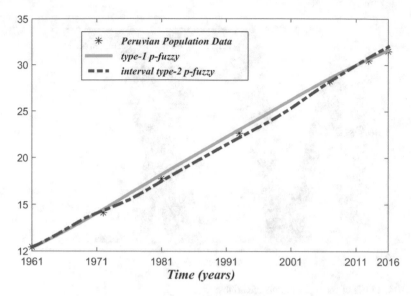

Fig. 3.45 Peruvian population data in millions and the trajectories from the p-fuzzy systems as a function of time

3.5.4 Conclusion

In this section, p-fuzzy systems have been used to obtain the trajectories for different population models. The decisions 1 and 2 of flowchart in Fig. 2.9 determine the type of FRBS to be used in the mathematical modeling. In the case of **Decision 2**, the type-1 p-fuzzy system for the Malthus model is based on the construction of a rule base and membership functions in the direction field of the ordinary differential equation associated with the model. In the case of **Decision 1**, the type-1 p-fuzzy system of the same model is trained by ANFIS based on the Malthusian principle that uses the proportionality constant and the interval with extremes given by the initial and final population.

The interval type-2 p-fuzzy system applied to the Malthus model has been built motivated by the type-1 FRBS. In both, the performance, with respect to the accuracy of the maximum of the relative error in each instant, has been a little more precise than using type-1 counterpart.

In addition, population data from Peru between the years 1961 and 2016 has been collected to show a different modeling technique. Thus, an approximation for $\frac{dp}{dt}$ is found, where p is the population, to define a type-1 FRBS, choosing **Decision 1**. Influenced by this FRBS, an interval type-2 p-fuzzy system is established. Again, when comparing via the maximum of the relative error at each instant, a small advantage of the interval type-2 fuzzy method compared to type-1 is observed.

Thus, the solution of the p-fuzzy system presents a great potential for modeling dynamic systems using FRBS of both types. More specifically, the potential to model autonomous differential equations is demonstrated in which the variation rate of state variables is modeled using fuzzy rules [24].

3.6   COVID-19 SIR Model Through Fuzzy Systems Identification of Type-1 and Interval Type-2

The World Health Organization (WHO) declared that a pneumonia outbreak caused by a new family member of coronaviruses, qualified as SARS-CoV-2 and named COVID-19, was detected in China at the end of December 2019. From there the virus has spread around the world, becoming a public health emergency of international concern. In fact, from the first records on December 31th 2019 up to July 31th 2020, there have been registered by the Worldometers [49] 5,916,688 confirmed infected and 682,392 confirmed deaths caused by COVID-19, even though real numbers might be much bigger. From the Worldometers source, a data set is collected to build a Susceptible-Infected-Removed (SIR) model. Being one of the first theoretical explanations of epidemics due to Kermack and McKendrick [29], this approach is often used to study the spread of infectious disease. The model is built tracking the number S of individuals susceptible to the disease, the number I of individuals infected with the disease, and the number R of individuals who have had the disease and are now either recovered or dead by the disease. The aim of this section is to develop a type-1 fuzzy SIR model from the methodology described in Sect. 2.1.4 and from there construct an interval type-2 fuzzy SIR model. The approach applied to obtain these models is based on an explicit identification function that describes the behavior of the variation rate of each fraction of SIR population with respect to the same fraction. From this relationship, the value of the SIR fraction at each instance using the Taylor's polynomial approximation is deduced [9]. The classic model obtained through empirical parameters, and initial conditions, is qualitative compared with the results of the fuzzy modeling.

3.6.1   The Coronavirus SARS-CoV-2 (COVID-19)

COVID-19 is a novel coronavirus with an outbreak of unusual viral pneumonia that has resulted in an epidemic of worldwide spread. The earliest reports of an illness caused by this type of virus were in the late 1920s, when an acute respiratory infection of domesticated chickens emerged in North America [16]. Human coronaviruses were first recognized in the 1960s [27]. Since then many other of these viruses have been identified, among which the most famous are the severe acute respiratory syndrome coronavirus, SARS-CoV in 2003, the Middle East respiratory syndrome, MERS-CoV in 2013, and the current SARS-CoV-2 in

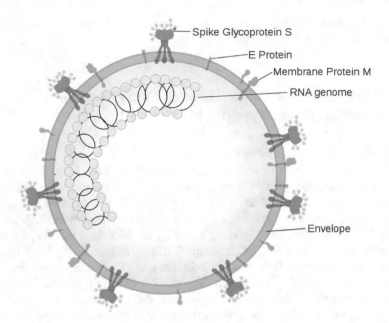

Fig. 3.46 COVID-19 structure. Source: Wikimedia Commons free media repository [2]

2019. These three viruses are from the family of virus named Betacoronaviruses [22, 30].

Depicted in Fig. 3.46 is the typical Betacoronaviruses cell structure, which contains structural proteins as: Spike Glycoprotein (S), Membrane Protein (M), Envelope Protein (E), and Nucleo Protein (N), inside of the Ribonucleic Acid (RNA) genome. The S protein is the essential factor of recognition for virus entrance and attachment to the host cells. The M protein is responsible in the formation of the virus, as well as the E protein. Checking the stability and protecting the RNA genome is the roll of the N protein. Moreover, proteins of the internal structural controlled some critical functions related to maintenance of the RNA genome and virus replications.

Many mathematical models have been developed to study preventive measures for COVID-19. These models can contribute in developing a plan to better distribute hospital beds and more appropriate medical care. In addition, these models can make predictions for the number of infected individuals, which would assist in the decision making of health-related institutions. These studies aim to present alternative measures such as: isolation policies; population mobility, locally or internationally; mandatory use of masks and alcohol use for hand hygiene; the prohibition of events involving a set number of individuals; among other options. As a result of these mathematical approaches, fewer individuals may be infected in a shorter period of time.

Fig. 3.47 The SIR dynamics

3.6.2 SIR Classic Model

This model describes the spread of a disease where no birth or death occurs during the period of analysis. Besides, an individual only can leave the susceptible group if they become infected, and the only way to leave the infected group is to recover or die. The other hypothesis assumed in this model is that those who have recovered or died are immune to the disease. Furthermore, the possibility of a susceptible individual to contract the disease at time t is proportional to the product $I(t) \cdot S(t)$. Considering the total population as a constant, N, and normalizing the values $s(t) = \frac{1}{N}S(t)$, $i(t) = \frac{1}{N}I(t)$, and $r(t) = \frac{1}{N}R(t)$, it is assumed that $s(t) + i(t) + r(t) = 1$, and as consequence

$$ds/dt + di/dt + dr/dt = 0, \quad t \geq 0. \tag{3.13}$$

Note that s, i, and r represent the fraction of the susceptible, infected, and removed population, respectively. Thus, a schematic of the dynamics is shown in Fig. 3.47.

These assumptions lead us to a system of three ordinary differential equations for $s(t)$, $i(t)$, and $r(t)$ as follows:

$$\begin{aligned} \frac{ds}{dt} &= -\kappa s i \\ \frac{di}{dt} &= \kappa s i - \tau i \\ \frac{dr}{dt} &= \tau i, \end{aligned} \tag{3.14}$$

where $\tau > 0$ is the recovery rate and $\kappa > 0$ is the transmission rate when an infected and a susceptible individual come in contact. Note that these expressions satisfy Eq. (3.13).

System (3.14) is solved using the fourth-order numerical Runge-Kutta method [9] for parameter values $\kappa = 0.12$, $\tau = 0.048$, initial conditions $s(0) = 0.9999999259$, $i(0) = 7.2 \cdot 10^{-8}$, and $r(0) = 2.1 \cdot 10^{-9}$, obtained empirically. In Fig. 3.48 the modeling result with the described characteristics is shown.

Note that t belongs to the range [0, 500], representing 500 days of the epidemic's model. The peak of the infected fraction, according to the modeling results, occurs in approximately 236 days.

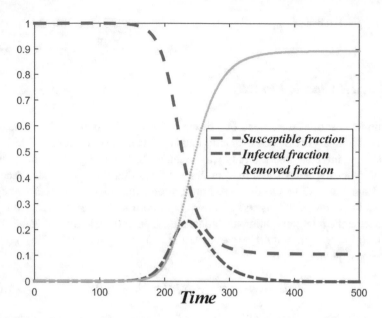

Fig. 3.48 Numerical solution for the SIR model

3.6.3   SIR Model Through Fuzzy System Identification of Type-1

From the Worldometers [49] source, a data set collected daily is organized as $EP = \{s(t_k), i(t_k), r(t_k)\}_{k=1}^{k=192}$, where t_1 is the day of the datum collected on December 31th 2019, and t_{192} is the one collected on July 31th 2020. These data are calculated throughout the collected data divided by the total world population, estimate in $N = 7,794,798,739$ at a specific date, considered under the hypothesis of the SIR model. The aim of this section is to provide functions $p^s(t)$, $p^i(t)$, and $p^r(t)$ that approximate the data $s(t_k)$, $i(t_k)$, and $r(t_k)$, $k = 1, 2, \ldots, 192$, respectively, as explained in the following.

Firstly, the general methodology which is used for each data set of EP is presented. Denoting $x(t_k)$, an element of EP, it is determined the matrix $D(k) = [x(t_k) \ \Delta x(t_k)]$, $\Delta x(t_k) = x(t_{k+1}) - x(t_k)$, $k = 1, 2, \ldots, 191$, and $D(192) = [x(t_{192}) \ \Delta x(t_{191})]$. Then, using the fuzzy system identification of Sect. 2.1.4.2, an explicit function $\delta^x(t) = \delta(x(t))$ that approximates $\{\Delta x(t_k)\}_{k=1}^{k=192}$ is obtained given by

$$\delta^x(t) = \frac{\sum_{j=1}^{c} \exp\left(-\frac{(x(t)-v_j)^2}{2\sigma_j^2}\right)(\theta_{j0} + \theta_{j1}x(t))}{\sum_{j=1}^{c} \exp\left(-\frac{(x(t)-v_j)^2}{2\sigma_j^2}\right)}, \tag{3.15}$$

where, c is the number of clusters, v_j are the cluster's centers of the clustering process, and σ_j is the standard deviation of the antecedent j in the TSK inference method , as well as the linear parameters $\theta_{j0}, \theta_{j1}, j = 1, 2, \ldots, c$.

Once the fuzzy system identification is applied to the data $D(k)$, $k = 1, 2, \ldots, 192$, at the first instant t_1, the Taylor's first degree polynomial approximation of $x(t_2)$, denoted by $p^x(t_2)$ is calculated, using the function $\delta^x(t)$ as follows:

$$p^x(t_2) = x(t_1) + \delta^x(t_1)(t_2 - t_1) = x(t_1) + \delta^x(t_1), \tag{3.16}$$

since $t_l - t_{l-1} = 1, l = 2, 3, \ldots, 192$. Likewise, the approximation for $x(t_3)$, is given by

$$p^x(t_3) = x(t_2) + \delta^x(t_2), \tag{3.17}$$

and substituting $x(t_2)$ for its approximation Eq. (3.16) in Eq. (3.17) is obtained

$$p^x(t_3) = x(t_1) + \delta^x(t_1) + \delta^x(t_2).$$

Proceeding inductively with the following instants of observation, the expression for the approximation of $x(t_l)$ is given by

$$p^x(t_l) = x(t_1) + \sum_{k=1}^{l-1} \delta^x(t_k), \ l = 2, 3, \ldots, 192. \tag{3.18}$$

The formula of Eq. (3.18) could be used as a calculation for intermediate values of t from the data collected. Indeed, let $t \in (t_{l-1}, t_l), l = 2, 3, \ldots, 192$. The value of $p^x(t)$ can be calculated as

$$p^x(t) = x(t_1) + \sum_{k=1}^{l-1} \delta^x(t_k) + \delta^x(t)(t - t_{l-1}). \tag{3.19}$$

For prediction or delayed information purposes, the calculation is the same using the future or delay values of $\delta^x(t)$. The precision of this procedure decreases as the instance increases in the future or decreases in the past. Notice that the functions $p^s(t)$, $p^i(t)$, and $p^r(t)$, are approximations for data $s(t_k)$, $i(t_k)$, and $r(t_k)$, $k = 1, 2, \ldots, 192$, respectively.

Recall that the parameters of the fuzzy identification modeling of type-1 are: β^x which determined the standard deviation of the antecedent of the inference method; α^x, which represents the projection level of the largest membership degrees, where x corresponds to each fraction of EP. These parameters for each fraction are: $\beta^s = 0.37$, $\alpha^s = 0.65$, $\beta^i = 0.35$, $\alpha^i = 0.6$, $\beta^r = 0.35$, and $\alpha^r = 0.6$. The clusters' centers v_j^x and the values of the parameters θ_{j1}^x and θ_{j0}^x are detailed in Table 3.18,

Table 3.18 Elements of function $\delta^x(t)$ of Eq. (3.15) corresponding to the $x(t_k)$ data

c	v_j^s	θ_{j1}^s	θ_{j0}^s	v_j^i	θ_{j1}^i	θ_{j0}^i	v_j^r	θ_{j1}^r	θ_{j0}^r
1	0.9999	$-0.7 \cdot 10^{-6}$	0	0.01	0	$-0.6 \cdot 10^{-3}$	0.006	0	$0.5 \cdot 10^{-5}$
2	0.9997	$-0.9 \cdot 10^{-5}$	0	0.18	0	0.0007	0.141	0	0.002
3	0.9994	$-0.1 \cdot 10^{-4}$	0	0.34	0	0.008	0.46	0	$0.4 \cdot 10^{-3}$
4	0.9991	$-0.2 \cdot 10^{-4}$	0	0.49	0	0.005	0.245	0	$-0.7 \cdot 10^{-3}$
5	0.9988	$-0.2 \cdot 10^{-4}$	0	0.59	0	-0.012	0.427	$0.2 \cdot 10^{-7}$	0
6	0.9985	$-0.3 \cdot 10^{-4}$	0	0.81	0	0.007	0.565	0	0.002
7	0.9979	$-0.3 \cdot 10^{-4}$	0	0.92	0	$0.2 \cdot 10^{-7}$	0.684	$0.2 \cdot 10^{-7}$	0
8	–	–	–	–	–	–	0.908	$0.2 \cdot 10^{-7}$	0

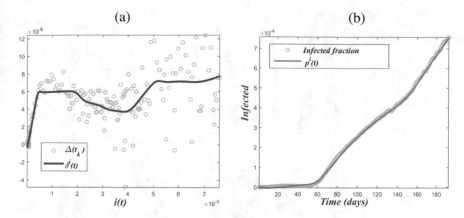

Fig. 3.49 The type-1 modeling's result for the $i(t_k)$ data. (**a**) $\delta^i(t)$ and $\Delta^i(t_k)$. (**b**) $p^i(t)$ and $i(t_k)$ data

where the superscript index s, i, and r corresponds to each element of the data set EP. For all simulations the value of *tol*, the stop criteria for the clustering process, is the same for the three fractions and is equal to 0.03. Finally, the total number of clusters for fractions s and i is 7, and 8 for r. Recall that all these parameters are obtained through an optimization process (see Sect. 2.1.4.2).

Shown in Fig. 3.49a is the function $\delta^i(t)$ of Eq. (3.15) along with $\Delta^i(t_k)$, $k = 1, 2, \ldots, 192$. In Fig. 3.49b is the function $p^i(t)$ of Eq. (3.19) and the $i(t_k)$, $k = 1, 2, \ldots, 192$. Note that, analyzing Fig. 3.49a, the infection rate, described by function $\delta^i(t)$, in the first days, rises abruptly and then remains steady with small variation. This can be seen as well in the real behavior of the epidemic in the world. Indicated in Fig. 3.49b is a qualitative relationship with the result of the classic model shown in Fig. 3.48 that the expected peak of the epidemic has not been reached.

3.6.4 SIR Model Through Fuzzy System Identification of Interval Type-2

Using the interval type-2 approach, the goal of this section is to provide functions $p_C^s(t)$, $p_C^i(t)$, and $p_C^r(t)$ representing the central value obtained through the interval type-2 modeling that approximate the data $s(t_k)$, $i(t_k)$, and $r(t_k)$, $k = 1, 2, \ldots, 192$, respectively.

From the construction of the type-1 fuzzy system identification, the parameters for the counterpart interval type-2 system have been extracted. In fact, the parameter α^x and the clusters' centers used in the type-1 fuzzy system identification are the same as shown in Table 3.1. Intervals that contain the parameters β^s, β^i, and β^r, are chosen empirically for the best results, more precisely, it is determined $\beta^s \in$

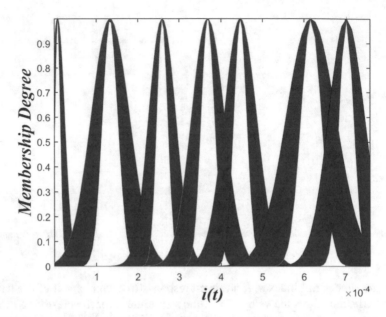

Fig. 3.50 Antecedents of the TSK inference method for $i(t_k)$ data

[0.18, 0.42], $\beta^i \in [0.16, 0.38]$, and $\beta^r \in [0.1, 0.5]$. The extreme values of these intervals are used to build the upper and lower membership functions of the interval type-2 inference's antecedents. An example of the set of antecedents built with this criterion is shown in Fig. 3.50, corresponding to the TSK inference method for data $i(t_k)$, $k = 1, 2, \ldots, 192$.

Likewise, the consequents of the TSK inference method are

$$Y^j = \left[\theta_{j0}^L, \ \theta_{j0}^R\right] + \left[\theta_{j1}^L, \ \theta_{j1}^R\right] x(t), \ \ j = 1, 2, \ldots, c.$$

The functions $\delta_L^x(t)$, $\delta_C^x(t)$, and $\delta_R^x(t)$ obtained by the interval type-2 fuzzy inference method are used to calculated the approximation value for the population using the formula of Eq. (3.15). As a consequence, the following values that approximate the data are acquired:

$$
\begin{aligned}
p_L^x(t_l) &= x(t_1) + \sum_{k=1}^{l-1} \delta_L^x(t_k), \\
p_C^x(t_l) &= x(t_1) + \sum_{k=1}^{l-1} \delta_C^x(t_k), \\
p_R^x(t_l) &= x(t_1) + \sum_{k=1}^{l-1} \delta_R^x(t_k), \ l = 2, 3, \ldots, 192.
\end{aligned}
\tag{3.20}
$$

As aforementioned in the type-1 modeling, the formula of Eq. (3.20) could be used as the calculation for intermediate values $t \in (t_{l-1}, t_l)$, $l = 2, 3, \ldots, 192$ as:

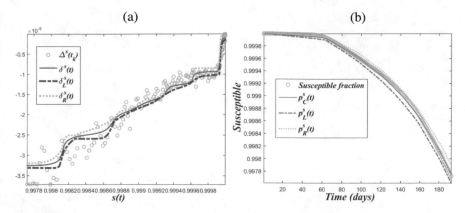

Fig. 3.51 The result for the $s(t_k)$ data corresponding to the interval type-2 modeling. (a) $\delta_L^s(t)$, $\delta_C^s(t)$, $\delta_R^s(t)$, and $\Delta^s(t_k)$. (b) $p_L^s(t)$, $p_C^s(t)$, $p_R^s(t)$, and $s(t_k)$ data

$$p_L^x(t) = x(t_1) + \sum_{k=1}^{l-1} \delta_L^x(t_k) + \delta_L^x(t)(t - t_l),$$
$$p_C^x(t) = x(t_1) + \sum_{k=1}^{l-1} \delta_C^x(t_k) + \delta_C^x(t)(t - t_l), \qquad (3.21)$$
$$p_R^x(t) = x(t_1) + \sum_{k=1}^{l-1} \delta_R^x(t_k) + \delta_R^x(t)(t - t_l).$$

Notice that the functions $p_C^s(t)$, $p_C^i(t)$, and $p_C^r(t)$ are approximations for the data $s(t_k)$, $i(t_k)$, and $r(t_k)$ $k = 1, 2, \ldots, 192$, respectively.

Depicted in Fig. 3.51 are (a) the functions $\delta_L^s(t)$, $\delta_C^s(t)$, and $\delta_R^s(t)$ along with $\Delta^s(t_k), k = 1, 2, \ldots, 192$. In Fig. 3.51b is shown the range determined by the values between $p_L^s(t)$ and $p_R^s(t)$ at each instant, resulted from the interval type-2 fuzzy system identification for the susceptible fraction $s(t_k)$, $k = 1, 2, \ldots, 192$.

Analyzing the behavior of the variation rate of susceptible fraction based on Fig. 3.51a it is observed that the function is negative, increasing, and, in the last days, has a significant growth, tending to zero. The curve of the susceptible fraction, $p_C^s(t)$, shown in Fig. 3.51b, is decreasing and is qualitatively equivalent to what occurs in the curve of the same fraction of the classic SIR graph of Fig. 3.48.

Shown in Fig. 3.52a are $\delta_L^i(t)$, $\delta_C^i(t)$, and $\delta_R^i(t)$ along with $\Delta^i(t_k)$, $k = 1, 2, \ldots, 192$, in the same graph. In Fig. 3.52b is shown the range of values between $p_L^i(t)$ and $p_R^i(t)$ for each t as a result of the interval type-2 approach for the infected fraction $i(t_k)$, $k = 1, 2, \ldots 192$.

Note that the function $p_C^i(t)$ shown in Fig. 3.52b has the same behavior as the curve $p^i(t)$ shown in Fig. 3.49b, in which case has been using the type-1 fuzzy system identification, both close to the data collected. In Fig. 3.52b, the function $p_R^i(t)$ could be representing the fraction of those infected in the alternative that most countries of the world had taken less preventive measures against COVID-19, such as social distancing, use of masks or alcohol gel for hand hygiene, and thoroughly washing hands. If most of the countries would require using these measures to

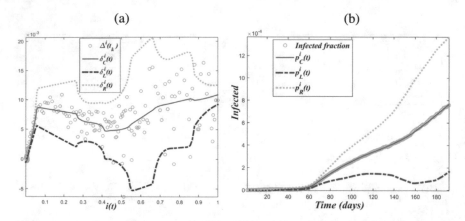

Fig. 3.52 The result of the interval type-2 modeling for the $i(t_k)$ data. (a) $\delta_L^i(t)$, $\delta_C^i(t)$, $\delta_R^i(t)$, and $\Delta^s(t_k)$. (b) $p_L^i(t)$, $p_C^i(t)$, $p_R^i(t)$, and $i(t_k)$ data

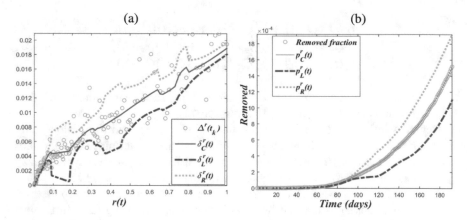

Fig. 3.53 The interval type-2 modeling's result for the $r(t_k)$ data. (a) $\delta_L^r(t)$, $\delta_C^r(t)$, $\delta_R^r(t)$, and $\Delta^s(t_k)$. (b) $p_L^r(t)$, $p_C^r(t)$, $p_R^r(t)$, and $r(t_k)$ data

contain the current coronavirus, then $p_L^i(t)$ seems to be an adequate model for the curve of the infected fraction in the world.

In Fig. 3.53a are the functions $\delta_L^r(t)$, $\delta_C^r(t)$, and $\delta_R^r(t)$, along with $\Delta^r(t_k)$, $k = 1, 2, \ldots, 192$, in the same graph. Shown in Fig. 3.53b is the range provide by the values between $p_L^r(t)$ and $p_R^r(t)$ at each instant, for the removed fraction $r(t_k)$, $k = 1, 2, \ldots, 192$.

Analyzing the curves $p_C^i(t)$ and $p_C^r(t)$ of Fig. 3.52b and Fig. 3.53b, respectively, it is observed that, in the last days, the curve $p_C^r(t)$ overcomes $p_C^i(t)$. This same behavior occurs with the infected and removed fraction in the classic model of Fig. 3.48. Thus, the model obtained by the interval type-2 fuzzy identification method and the classic model are qualitatively equivalent.

Table 3.19 Comparison of the error between the type-1 and interval type-2 fuzzy systems for identification

Fuzzy system identification	Susceptible	Infected	Removed
Type-1	$5.3488 \cdot 10^{-5}$	$1.5473 \cdot 10^{-5}$	$1.2632 \cdot 10^{-5}$
Interval type-2	$4.4446 \cdot 10^{-5}$	$1.2217 \cdot 10^{-5}$	$1.1542 \cdot 10^{-5}$

The results of the fuzzy system identification for both types are compared using as measure of accuracy the maximum absolute value at each instant. Detailed in Table 3.19 are the results of the comparison.

3.6.5 Conclusion

The classical Susceptible-Infected-Removed model is qualitatively compared with two fuzzy models of type-1 and interval type-2, the latter based on the information of the former. Both fuzzy models return a qualitative similar response that matches with the classical model and with the data behavior extracted from the world confirmed numbers. For each of the population fractions studied, three polynomials are obtained from the interval type-2 modeling. One central polynomial that approximates the collected data with high precision and the other two determine a range of uncertainty that can be interpreted as the imprecision existing in the practice of data collection.

Furthermore, a comparison between the fuzzy approaches is made throughout the maximum absolute error, concluding a slightly better performance of the interval type-2 fuzzy modeling over the type-1 counterpart.

3.7 Summary

Six mathematical models of biological phenomena as applications of the interval type-2 fuzzy theory have been developed in this chapter. A succinct description of each application is presented in the following.

- The first model developed, originated from a pharmacological phenomenon, describes the elimination of drugs from the individual's body. The data of urinary volume, creatinine clearance, and blood pH of three individuals are used to calculate the velocity of the drug elimination using type-1 and interval type-2 Fuzzy Rule-Based System (FRBS).
- The goal of the second application is to confirm that an interval type-2 FRBS can present a wider range of possibilities to identify patients with organ-confined prostate cancer, meanwhile the type-1 fuzzy approach shows only one output.

- In the third application, p-fuzzy systems are used as an alternative methodology to model the transference of Human Immunodeficiency Virus (HIV)-seropositive population fraction from asymptomatic to symptomatic. An interval type-2 p-fuzzy system is developed, obtaining a range that includes the type-1 p-fuzzy system approximation and the Peterman's data numerically integrated, which motivates the deterministic model.
- Presented in the fourth section is a study of two systems of ordinary differential equations for HIV dynamics. Both models are used to construct two Cellular Automata (CA), where their inputs come from two interval type-2 FRBS. The CA's outputs return qualitative features that validate the mathematical modeling.
- The aim of the fifth section is to introduce three different procedures using p-fuzzy systems to obtain approximations for the population models trajectories. The classic Malthus model is analyzed in two cases. One from a previous built type-1 FRBS using Mamdani inference method along with a centroid defuzzification technique. The second one through a trained Adaptive Neuro-Fuzzy Inference Systems (ANFIS) and utilizing the Malthusian principle, to obtain the type-1 FRBS. The third model is built from a Peruvian population collected data using as well the ANFIS technique. All the type-1 FRBS motivate the development of the interval type-2 fuzzy counterpart.
- The last model presented in this chapter refers to the world international concern of the year 2020: the epidemic cause by COVID-19. A Susceptible-Infected-Removed model is explored to obtain interval type-2 outputs that could be used to understand the uncertainties of the epidemics' confirmed numbers of susceptible-infected-removed individuals. The outputs are obtained utilizing a fuzzy identification technique that result in explicit expressions for the approximations of each fraction at any instant.

References

1. AIDSINFO: Offering information on HIV/AIDS treatment, prevention, and research. https://aidsinfo.nih.gov/about-us. Accessed July 2020
2. Altaf, S.: Wikimedia commons free media repository. Own work, CC BY-SA 4.0. https://commons.wikimedia.org/w/index.php?curid=88349537. Accessed September 2020
3. Anderson, R., Medley, G.F., May, R.M., Johnson, A.M.: A preliminary study of the transmission dynamics of the human immunodeficiency virus HIV, the causative agent of AIDS. IMA J. Math. Appl. Med. Biol. **3**, 229–263 (1986)
4. Atkinson, K.E.: An Introduction to Numerical Analysis, 2nd edn. John Wiley and Sons, New York (1989)
5. AVERT: HIV time. https://timeline.avert.org/. Accessed July 2020
6. AVERT: https://www.avert.org/professionals/history-hiv-aids/overview#footnote37_spskps5. Accessed September 2020
7. Barros, L.C., Bassanezi, R.C.: Tópicos de Lógica Fuzzy e Biomatemática, vol. 5, 2nd edn. Coleção MECC - Textos Didáticos, Campinas (in Portuguese) (2010)
8. Bassanezi, R.C., Ferreira Jr, W.C.: Equações diferenciais com aplicações. Ed. Harbra, S. Paulo (1978, in Portuguese)
9. Burden, R.L., Faires, J.D.: Numerical Analysis, 9th edn. Brooks/Cole Cengage Learning, Boston (2011)

10. Castanho, M.J.P.: Construção e avaliação de um modelo matemático para predizer a evolução do câncer de próstata e descrever seu crescimento utilizando a teoria dos conjuntos fuzzy. Ph.D. Thesis, FEEC – UNICAMP, Campinas (2005, in Portuguese)
11. Castro, J.R., Castillo, O., Martínez, L.G.: Interval type-2 fuzzy logic toolbox. Eng. Lett. **15**(1) (2007)
12. Castanho, M.J.P., Hernandes, F., Ré, A.M.D., Rautenberg, S., Billis, A.: Fuzzy expert system for predicting pathological stage of prostate cancer. Expert Syst. Appl. **40**(2), 466–470 (2013)
13. Castanho, M.J.P., Jafelice, R.S.M., Lodwick, W.: Interval type-2 fuzzy system in prostate cancer. In: Conferência Brasileira de Dinâmica, Controle e Aplicações - DINCON2017. São José do Rio Preto (2017)
14. Coutinho, F.A.B., Lopez, L.F., Burattini, M.N., Massad, E.: Modelling the natural history of HIV infection in individuals and its epidemiological implications. Bull. Math. Biol. **63**, 1041–1062 (2001)
15. Ermentrout, G., Edelstein-Keshet, L.: Cellular automata approaches to biological modeling. J. Theor. Biol. **160**, 97–133 (1993)
16. Estola, T.: Coronaviruses, a new group of animal RNA viruses. Avian Dis. **14**(2), 330–336 (1970)
17. Ferreira, D.P.L.: Sistema p-fuzzy aplicado às equações diferenciais parciais. Master's thesis, FAMAT – UFU, Uberlândia/MG (in Portuguese) (2012)
18. Ferreira, G.S., Jafelice, R.S.M.: Modelo p-fuzzy do tipo 2 para a transferência da população HIV assintomática para sintomática. Technical report, Universidade Federal de Uberlândia (in Portuguese) (2018)
19. Hanley, J.A., McNeil, B.J.: The meaning and use of the area under a receiver operating characteristic(ROC) curve. Radiology **143**, 29–36 (1982)
20. Hohenwarter, M.: Geogebra 5.0. http://www.geogebra.org (2020). Accessed September 2020
21. Holland, J.H.: Adaptation in Natural and Artificial Systems: An Introductory Analysis with Applications to Biology, Control, and Artificial Intelligence. University of Michigan Press, Michigan (1975)
22. ISID: The continuing 2019-ncov epidemic threat of novel coronaviruses to global health: The latest 2019 novel coronavirus outbreak in Wuhan, China. Int. J. Inf. Dis. **91**, 264–266 (2020)
23. Jafelice, R.M.: Modelagem fuzzy para dinâmica de transferência de soropositivos para HIV em doença plenamente manifesta. Ph.D. Thesis, FEEC, Universidade Estadual de Campinas, Campinas (2003, in Portuguese)
24. Jafelice, R.M., Barros, L.C., Bassanezi, R.C.: Teoria dos Conjuntos Fuzzy, vol. 17, 2nd edn. SBMAC, São Carlos (2012, in Portuguese)
25. Jafelice, R.S.M., Cabrera, N.V., Câmara, M.A.: Sistemas p-fuzzy utilizando conjuntos fuzzy do tipo 2 intervalar. Biomatemática - UNICAMP **28**, 1–14 (2018, in Portuguese)
26. Jafelice, R.S.M., Pereira, B.L., Bertone, A.M.A., Barros, L.C.: An epidemiological model for HIV infection in a population using type-2 fuzzy sets and cellular automaton. Comput. Appl. Math. **38**(141) (2019)
27. Kahn, J.S., McIntosh, K.: History and recent advances in coronavirus discovery. Pediatr. Infect. Dis. J. **24**(11), 223–227 (2005)
28. Kattan, M.W., Eastham, J.A., Wheeler, T.M., Maru, N., Scardino, P., Erbersdobler, A., Graefen, M., Huland, H., Koh, H., Shariat, S., Slawin, K., Ohori, M.: Counseling men with prostate cancer: a nomogram for predicting the presence of small, moderately differentiated, confined tumors. J. Urol. **170**, 1792–1797 (2003)
29. Kermack, W.O., McKendrick, A.G.: Contributions to the mathematical theory of epidemics. J. R. Stat. Soc. **115**, 700–721 (1927)
30. Lalchhandama, K.: A biography of coronaviruses from IBV to SARS-CoV-2, with their evolutionary paradigms and pharmacological challenges. Int. J. Res. Pharm. Sci. **11**, 208–218 (2020)

31. Lopes, W.A., Jafelice, R.S.M.: Fuzzy modeling in the elimination of drugs. In: Proceedings of the 2005 International Symposium on Mathematical and Computational Biology, vol. 3, pp. 39–355. International Symposium on Mathematical and Computational Biology, E–papers Serviços Editoriais Ltda, Petrópolis (2006)
32. Malthus, T.R.: An Essay on the Principle of Population. J. Johnson, London (1798)
33. Mendel, J.M.: Type-2 fuzzy sets and systems: an overview. IEEE Comput. Intell. Magaz. **2**(1), 20–29 (2007). https://doi.org/10.1109/MCI.2007.380672
34. Moore, E.F.: Machine models of self-reproduction. In: Proceedings of Symposia in Applied Mathematics, vol. XIV, pp. 17–34 (1962)
35. O'Brien, T., Shaffer, N., Jaffe, H.: Transmissão e infecção do vírus HIV. In: M. Sande, P. Volberding (eds.) Tratamento Clínico da AIDS, 3rd edn., pp. 3–14. Revinter, Stuttgart (1995, in Portuguese)
36. Ozkan, I.A., Saritas, I., Sert, U.: A type-2 fuzzy expert system application for the determination of prostate cancer risk. In: Proceedings of International Conference on Advanced Technology and Sciences (ICAT'15), pp. 498–501 (2015)
37. Pereira, B.L., Jafelice, R.S.M.: Estudo da propagação do HIV em uma população baseado em autômatos celulares. Technical Report. Universidade Federal de Uberlândia (2013, in Portuguese)
38. Perelson, A., Nelson, P.: Mathematical analysis of HIV-1 dynamics in vivo. SIAM Rev. **41**, 3–44 (1999)
39. Peterman, T.A., Drotman, D.P., Curran, J.W.: Epidemiology of the acquired immunodeficiency syndrome AIDS. Epidemiol. Rev. **7**, 7–21 (1985)
40. Ralescu, D., Y.Ogura, S.Li: Set defuzzification e choquet integral. Int. J. Uncertainty Fuzziness Knowledge Based Syst. **9**, 1–12 (2001)
41. Ren-Jieh, K., Man-Hsin, H., Wei-Che, C., Chih-Chieh, L., Yung-Hung, W.: Application of a two-stage fuzzy neural network to a prostate cancer prognosis system. Artif. Intell. Med. **63**(2), 119–133 (2015)
42. Renning, C.: Collective behaviour: Emergent dynamics in populations of interacting agents. In: Seminar Artificial Life (1999/2000)
43. Saag, M.: Diagnóstico laboratorial da AIDS presente e futuro. In: M. Sande, P.A.Volberding (eds.) Tratamento Clínico da AIDS, 3rd edn., pp. 27–43. Revinter, Stuttgart (1995, in Portuguese)
44. Silva, C., Jafelice, R.: Uso de autômato celular em tratamento dos HIV positivos dependendo da evolução da doença. In: 18^0 SIICUSP. São Paulo (2010, in Portuguese)
45. Splettstoesser, T.: Scistyle. www.scistyle.com. Accessed August 2020
46. Steyerberg, E., Roobol, M., Kattan, M., der Kwast, T.V., de Koning, H., Schröder, F.: Prediction of indolent prostate cancer: validation and updating of a prognostic nomogram. J. Urol. **177**(1), 107–112 (2007)
47. UNAIDS: http://www.unaids.org (2020). Accessed August 2020
48. Villela, M.F.S., Freitas, K.B., Jafelice, R.S.M.: Modelos p - fuzzy de transferência da população HIV assintomática para sintomática. In: Anais do XIV do Congresso Latino – Americano de Biomatemática. Campinas (2007, in Portuguese)
49. Woldometers: COVID-19 coronavirus pandemic. https://www.worldometers.info/coronavirus. Accessed 14 August 2020
50. Zanini, A.C., Seizi, O.: Farmacologia Aplicada. Atheneu, Brazil (1994, in Portuguese)

Chapter 4
Interval Type-2 Fuzzy Sets in the Future: Scientific Projects for Development

The applications described in Chap. 3 are related to topics in the biological field, such as drug elimination, prostate cancer-related diseases, population growth, and epidemiological disease caused by HIV and COVID-19. In this chapter some scientific projects are suggested that may be useful for the development of research using an interval type-2 fuzzy set.

4.1 Project: Multicompartmental Pharmacokinetic Models Through Takagi–Sugeno–Kang Inference Method

Presented in this section is a proposal for a project to be developed using interval type-2 fuzzy sets based on the model developed by Menegotto [5] and also published in [2]. In Menegotto [5] a decay model is obtained for the curve of the concentration of drugs through a type-1 FRBS using the TSK inference method. The type-1 FRBS consists of two fuzzy rules, whereas the set of data fits into a two-compartment model that has been extracted from [3]. Therefore, two phases are analyzed: the drugs' distribution (Phase 1) and the elimination phase (Phase 2) [8].

For the type-1 FRBS, the input variable is time in hours (t) comprised of the linguistic terms: Low (L) and Non-Low (NL). The input variable membership functions that are trapezoidal are presented in Fig. 4.1. The output of each fuzzy rule is the plasma concentration of the drug, which, on a semi-logarithmic scale, is given by the lines of the Phase 1 and Phase 2 that are determined by the least-squares method. The straight lines of the each phase are:

- Phase 1: $Y^1 = 0.11394 - 0.0751t$,
- Phase 2: $Y^2 = -0.08618 - 0.003995t$.

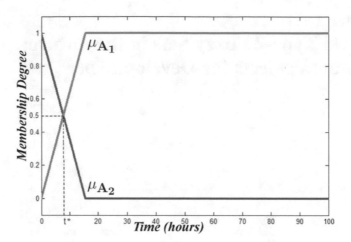

Fig. 4.1 Input membership functions

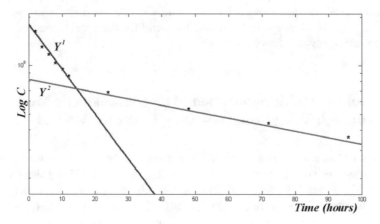

Fig. 4.2 Semi-logarithmic scale output lines

In Fig. 4.2 the experimental data and the lines corresponding to each phase, Y^1 and Y^2, on semi-logarithmic scale are presented.

As the lines in Fig. 4.2 are in semi-logarithmic scale, namely, $Y^1 = \log_{10} C^1$ and $Y^2 = \log_{10} C^2$, C^1 and C^2 being drugs' concentration functions depending on time, then $C^1 = 10^{Y^1}$ and $C^2 = 10^{Y^2}$. The rules base is given by:

- R^1: If t is L, then $C^1 = 10^{Y^1}$ (Phase 1);
- R^2: If t is NL, then $C^2 = 10^{Y^2}$ (Phase 2).

The general output of the system using the TSK inference's method is the curve given by

$$C(t) = \frac{\mu_{A_1}(t) \cdot 10^{Y^1} + \mu_{A_2}(t) \cdot 10^{Y^2}}{\mu_{A_1}(t) + \mu_{A_2}(t)}, \tag{4.1}$$

where $C(t)$ is the concentration of the drug in the individual's blood and the weights (defined by the minimum norm operator) are obtained by fuzzy numbers A_1 and A_2. This curve is qualitatively compared with the deterministic solution resulting from the resolution of an ordinary differential equation system associated with the model's two phases: distribution and elimination of the drugs. Both curves obtained by the different modeling are consistent with the pharmacology information.

The proposal is to build an interval type-2 FRBS, where the FOUs of the input and output variables contain the input and output membership functions of the type-1 FRBS, respectively. A suggestion for the finalized project is to compare the results of the different approaches, deterministic, and type-1 and interval type-2 FRBS, as made in [5]. In this way, the development of this project through the interval type-2 FRBS can improve the model by expanding the uncertainty present in the decay of the drugs in each individual's body.

4.2   Project: Prostate Cancer Diagnostic Model

The research project's suggestion is to build an interval type-2 FRBS that uses the risk of prostate cancer as the output variable. This study is important since, for example, in Brazil prostate cancer is the second most common among men (behind only non-melanoma skin cancer). In absolute values and considering both diseases, it is the second most common type. The incidence rate is higher in developed countries compared to developing ones [6].

Thus, it is initially necessary to collect data from individuals who have undergone prostate cancer screening tests, with the help of a specialist in the field or information from other registered studies presented in the literature. Performing this first stage and applying the ANFIS routine to determine a type-1 FRBS, the input variables could be: the individual's age, clinical stage, PSA level or its derivatives (PSA density, PSA speed, expected PSA at which density and expected PSA depend on the prostate volume), and dimensions of the prostate, where these data are obtained by image examination procedure. The output variable is the risk of prostate cancer.

This methodology may contribute to a possible diagnosis of this cancer. Recall that the Area Under the Curve of the Receiver Operating Characteristic, in the case of the interval type-2 fuzzy sets result, is an interval that includes the type-1 output (see Sect. 3.2.1). Therefore, the interval type-2 fuzzy approach has expanded the possibilities of identifying individuals with the characteristics associated with this disease. For this reason, an interval type-2 FRBS could be built with the input and output variables suggested for the type-1 FRBS.

4.3 Project: Adding a Fuzzy Parameter in Cellular Automaton of the HIV-Seropositive Population Dynamics with Antiretroviral Therapy

The HIV-seropositive evolution model with antiretroviral treatment [1] has been presented in Sect. 3.4. This research proposal is to change the system of ordinary differential equations (3.9) considering the parameter d, percentage of mortality of the fraction of symptomatic individuals due to the AIDS virus, as the output of an interval type-2 FRBS. The modified system with the parameter added is described by

$$\frac{da}{dt} = -\lambda(v, c, f)\, a + \gamma(v, c, f)\, s - \mu a \qquad\qquad a(0) = a_0, \quad 0 < a_0 < 1$$
$$\frac{ds}{dt} = \lambda(v, c, f)\, a - \gamma(v, c, f)\, s - (\mu + d(v, c, f))\, s \quad s(0) = s_0 = 1 - a_0$$
$$a + s = 1,$$

$$(4.2)$$

where the variables and parameters involved are as follows: a is the fraction of asymptomatic individuals, as a function of time; s is the fraction of symptomatic individuals, as a function of time; λ is the transference rate of the fraction of the population from asymptomatic to symptomatic depending on the viral load, v, the $CD4+$ T lymphocyte level, c, and the adherence to the treatment, that is, follow the medication administration protocol, f; γ is the return rate from the symptomatic to asymptomatic fraction of the population, which depends on v, c, and f; and μ is the percentage of mortality from natural causes of the two fractions of individuals. Initially, it is suggested to build the interval type-2 FRBS with input variables v, c, and f, where the membership functions are constructed in a similar way to the graphs presented in Sect. 3.4. The output variable d, percentage of mortality of the fraction of symptomatic individuals due to the AIDS virus, with support given, for instance, by [0, 0.1]. A second step would be to determine the solution of the system (4.2). In this project, it would also be interesting to simulate the CA that takes into account the dynamics of the system (4.2). In this CA asymptomatic and symptomatic HIV-seropositive individuals would interact artificially and would have three FRBS feeding its inputs through the output of interval type-2 FRBS. It is believed that this project will include uncertainty in another important parameter, since the mortality of symptomatic individuals due to AIDS is directly linked to the viral load, the $CD4+$ T lymphocyte level, and the adherence to treatment.

4.4 Project: Verhulst's p-Fuzzy System

Another suggested project is to carry out, for undergraduate students or readers interested in the subject, is to extend the Verhulst's p-fuzzy model presented in

Sect. 2.1.3 to an interval type-2 FRBS. This would be a good exercise as a means of learning, in a solid way, the theory of interval type-2 fuzzy sets. Recall that the p-fuzzy modeling obtained with type-1 FRBS in Sect. 2.1.3 compared with the analytic solution has a similar behavior, qualitative and quantitative manner. Motivated by the results of other p-fuzzy population models presented in this book, it is believed that the interval type-2 FRBS may contribute to a better understanding of the phenomenon of population growth.

4.5   Project: Predator–Prey p-Fuzzy System

Among the models of interaction between species, the classic predator–prey model is highlighted, where the mathematical formulation is comprised of the Malthusian model (exponential growth or decrease) and the law of mass action (interaction between species). The predator–prey model , also known as Lotka–Volterra model [4, 9], has been a starting point for the development of new techniques and mathematical theories. The model is proposed by the equations:

$$\begin{cases} \dfrac{dN_1}{dt} = a_{11}N_1 - a_{12}N_1N_2 \\ \dfrac{dN_2}{dt} = -a_{21}N_2 + a_{22}N_1N_2 \end{cases} \tag{4.3}$$

where $N_1 = N_1(t)$, the prey's population density, $N_2 = N_2(t)$, the predator's population density both at each time t; a_{11} is the prey's growth rate; a_{12} is the probability for a predator to kill a prey at each encounter; a_{21} is the death rate of predator in the absence of preys; and a_{22} denotes the rate of conversion from prey to predator, all positive constants.

The authors of [7] present the p-fuzzy system for the predator–prey model using type-1 FRBS with a Mamdani inference method and center of gravity as defuzzification technique. The project would be similar to the p-fuzzy system through an interval type-2 FRBS with the same inference method. This suggestion would be interesting as this book does not present a p-fuzzy system comprised of two equations.

The authors believe that the projects can be carried out by readers with

"a dedication *moderately* accentuated, but *potentially* beneficial."

References

1. Jafelice, R.M.: Modelagem fuzzy para dinâmica de transferência de soropositivos para HIV em doença plenamente manifesta. Ph.D. Thesis, FEEC – Universidade Estadual de Campinas, Campinas (2003, in Portuguese)
2. Jafelice, R.M., Barros, L.C., Bassanezi, R.C.: Teoria dos Conjuntos Fuzzy, vol. 17, 2nd edn. SBMAC, São Carlos (2012, in Portuguese)
3. Jamili, F.: Clinical pharmacokinetics of selected classes of drugs: pharmacokinetic compartments. http://www.pharmacy.ualberta.ca/pharm415pharmaco.htm. Accessed 2011
4. Lotka, A.J.: Elements of Physical Biology. Williams & Wilkins, Philadelphia (1925)
5. Menegotto, J.: Aplicação da teoria dos conjuntos fuzzy em modelos farmacocinéticos multicompartimentais. Master's Thesis, IMECC, Universidade Estadual de Campinas, Campinas (2011, in Portuguese)
6. MH: Brazilian Ministry of Health: Câncer de próstata. https://www.inca.gov.br/tipos-de-cancer/cancer-de-prostata (in Portuguese). Accessed August 2020
7. Peixoto, M.S., Barros, L.C., Bassanezi, R.C.: Predator–prey fuzzy model. Ecol. Model. **214**(1), 39–44 (2008)
8. Valle, L., Oliveira, F., Luca, R.D., Olga, S.: Farmacologia integrada e farmacologia básica, 1st edn. Editora Atheneu, Sao Paulo (1988, in Portuguese)
9. Volterra, V.: Theory of Functionals and of Integral and Integro-differential Equations. Blackie & Son, Glasgow (1930)

Index

© The Author(s), under exclusive license to Springer Nature Switzerland AG 2021
R. S. da Motta Jafelice, A. M. A. Bertone, *Biological Models via Interval Type-2
Fuzzy Sets*, SpringerBriefs in Mathematics,
https://doi.org/10.1007/978-3-030-64530-4

Printed in the United States
By Bookmasters